The Enlightened Savage

Using Primal Instincts for Personal & Business Success

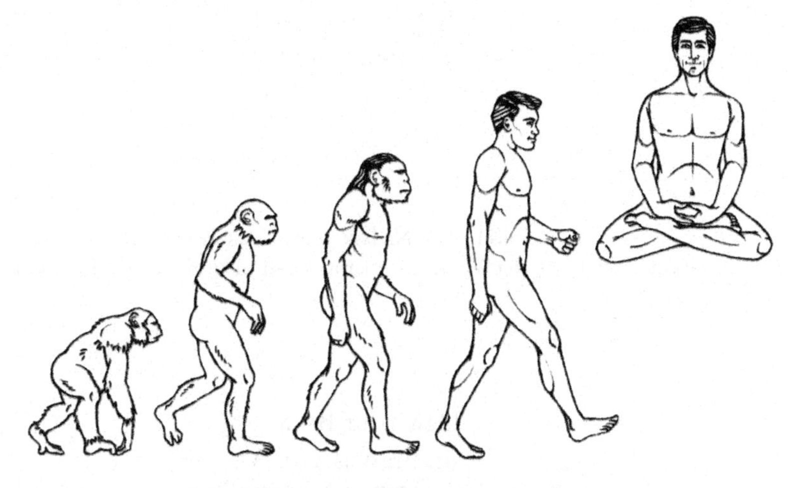

Anthony Hernandez

Second Edition

The Enlightened Savage

Using Primal Instincts for Personal & Business Success

Second Edition

Anthony Hernandez

Illustrations by Lori Nadaskay (lnadaskay@aol.com)
Cover photo by Martin Wierzbicki (martin.wierzbicki@photosbymartin.com

Dawnstar Books
San Francisco, CA
(415)786-2081
www.DawnStarBooks.com
info@DawnStarBooks.com

Hernandez, Anthony 1968-
The Enlightened Savage/Anthony Hernandez
ISBN: 978-0-9855793-3-3 (paperback)
ISBN: 978-0-9855793-4-0 (Amazon Kindle)

Advance Praise for *The Enlightened Savage*

"Anthony Hernandez's book is a motivational book in the deepest sense—-a vivid exploration of the elemental and instinctual ways in which our modern minds are wired by evolution. It's a book that explains a *lot* about why we do the things we do and, with that understanding, how we can do them with more graceful and efficient results."

Gregg Levoy
Author, *Callings: Finding and Following an Authentic Life*
www.gregglevoy.com

"Anyone who recognizes the value of increasing their self-awareness to increase their success with ease must read *The Enlightened Savage.* Anthony Hernandez combines a solid understanding of how people operate with effective strategies to shed limiting beliefs, find your life's calling, and begin living the life of abundance and joy that you desire."

Caterina Rando, MA, MCC
Keynote speaker, success coach
Author, *Learn to Power Think*
www.caterinar.com

"Conscious awareness of how our brains and bodies are 'hard-wired' for specific emotional responses to certain stimuli is critical to gaining true emotional intelligence, and becoming the conscious creators of our lives that we are meant to be. Anthony Hernandez is on the cutting edge of this essential knowledge with *The Enlightened Savage.* A must read for anyone who desires to live a personally empowered, awakened and aware life."

Ilana Moss
Author, *The Heartcompass Owner's Manual*

"*The Enlightened Savage* is a wonderfully clear and entertaining exploration of our psychological roots and our power to create the life we want. The book is filled with practical, real world exercises to move the reader step-by-step from being hijacked by repetitive patterns to being joyous, empowered and free. How many times have we read books with wise concepts that never translate into how we are in our daily lives? This book is filled with ways to help the reader bring these insights to life in the real world. Anthony's enthusiasm and personal success with this process is clear throughout the book; he has created an intelligent and immensely practical journey into our possibilities."

Paul Clark
Director of Coaching
Conversations with God Coaching Services
www.cwg.org

"Want to tap into your inner self? In *The Enlightened Savage*, Anthony Hernandez moves you to tap into your inner strengths and discover your primal inner self. A powerful read that is connects to your inner core."

Mary Hambledon
Soul Canyon
www.soulcanyon.com

"Often controversial, always provocative, Anthony Hernandez charts a unique course for personal improvement that combines a deep understanding of the human psyche, a respect for core values, and the crucial acknowledgement that work and success should be fun."

Shel Horowitz
Originator of the Business Ethics Pledge
Author, *Principled Profit: Marketing That Puts People First*
www.principledprofits.com

"For those of you wanting to open, expand and finally change your life, but don't know where to begin, *The Enlightened Savage* is the book for you! Anthony has done a wonderful job of taking you on a journey that will lead you to an understanding of self in an easy and light-hearted way. This book is filled with a depth of information, insights and wisdom that will help you to envision and create the life you want.

This is definitely a guy book, a great way for the guy in your life to begin to open up. So gals, when you finish reading it, pass it right on!"

Glenda Feinsmith LCSW
National Board of Clinical Hypnotherapists Diplomat

"Take off your mask and be real. Anthony Hernandez teaches us clearly and concisely how taking off our mask is the key to tapping into our primal power. A must read for anyone who wants to get real."

Rob Hambledon
Founder of Soul Canyon
www.soulcanyon.com

"*The Enlightened Savage* is more than just a great book, it is an empowering program that entertains as it asks you to engage in an exciting personal exploration. Whether you agree with all the concepts in this book or not, no one can deny that it attempts to answer some of the most intriguing and impactful questions ever posed."

Rachael Kennedy
Director of Education
Conversations with God Coaching Services
www.cwg.org

"As a scientist and CEO, I can appreciate the quest to reveal the invisible. Anthony's book makes a huge leap forward in explaining some core scientific principles that drive our behavior. Understanding the core mechanics of our subconscious allows us to eliminate destructive patterns and take control of our lives. It's about time that somebody wrote this book."

Bryan Brandenburg
CEO & Chief Science Officer
Zygote Media Group and 3DScience.com
www.zygote.com and www.3dsci.com

"As a real estate investor, artist, and consultant, I am continually looking for ways to stay motivated and creative. Anthony Hernandez's book, *The Enlightened Savage*, is a winner! It is well written and contains a twist that sets it apart from other books and makes it a must read for anyone looking to better themselves and improve their lives. While reading, I realized that this isn't just another self-improvement book but a 'self-fulfillment' book. If you've ever been stuck or at a place in life where you're sick and tired of the status quo, then you owe it to yourself to read this book.

The Enlightened Savage is unique because it tells the complete story. Mr. Hernandez provides simple comprehensive steps to a better more fulfilling life, gives us the reasons behind our own human nature (in other words, why we do the things we do, then shows us how to use our human instincts to our advantage. Most self-improvement books only deliver half the story but Mr. Hernandez delivers both sides of the coin, the how-to and the very important 'why?' You need not believe in the theory of evolution to get all that this book has to offer and become an Enlightened Savage. This book takes you on a journey that can lead you to the next level on your path to personal success and fulfillment. I look forward to reading many more books from Mr. Hernandez in the future!"

Darla Anderson
President, DAPA, Inc.

www.darlaanderson.com
www.blueorchidgallery.com

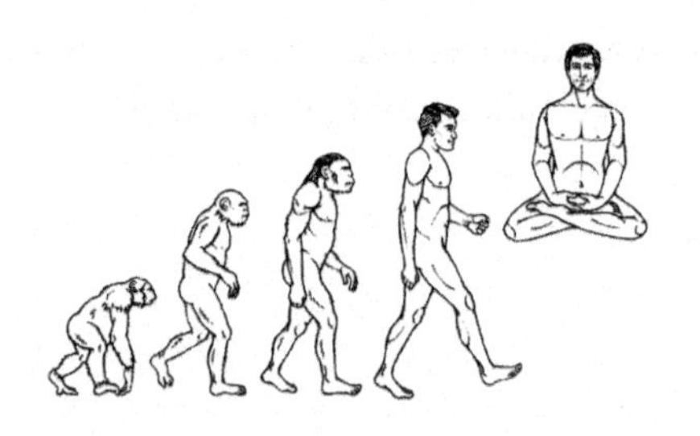

Table of Contents

Dedication

This book is dedicated to my son Logan, who hugged me the moment he learned to walk and who hugs me still.

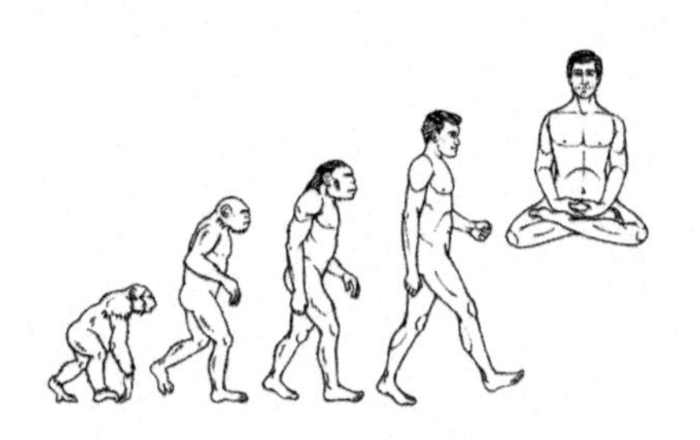

Acknowledgements

The Enlightened Savage would have been absolutely impossible without much help and many contributions from many amazing people, and I'd like to take a few pages to thank them.

This book has its roots in the spring of 2002, when I created two video products designed to help authors publish and market their books more effectively. Rose Wahlin donated many hours of her time helping me film and produce those first products. Thanks to those efforts, I met Jay Conrad Levinson, the father of Guerrilla Marketing, his wife Jeannie, and his daughter Amy.

The Levinsons took me under their wing and have been mentoring and training me in the ways of Guerrilla Marketing, helping me launch my own business and get more involved with Guerrilla Marketing International. They also introduced me to Larry Loebig and Rick Eggers, founders of the Guerrilla Marketing Business Academy, where I am obtaining my Masters degree in Psychology with a focus in Guerrilla Marketing through Western Institute for Social Research. Some of the material in this book is inspired by several of Jay's *Guerrilla Marketing* books.

Larry Loebig is a coach, professor, and mentor. He is constantly pushing me to be better than my best and providing constant support in an environment that encourages me to think freely. At a time when I was questioning the viability of this project, he encouraged me to stay the course and finish writing.

Caterina Rando's *Success With Ease* program worked miracles, helping me see beyond the effort I was putting into my endeavors and look to the results. That one change has brought about tremendous change in my life and business. She graciously allowed me to include some of her material in this book.

T. Harv Eker's 3-day *Millionaire Mind* seminar was my first foray into motivational seminars, and I loved every second. His teachings and energy inspired me and helped crystallize the ideas that became *The Enlightened Savage.*

As I was writing early versions of this work, I was searching hard for effective methods of changing mental programming.

I found Dr. Morty Lefkoe's site, tested and researched his processes, and loved what I saw and experienced. Dr. Lefkoe contributed the processes you'll find later in this book.

I owe a huge debt of gratitude to my friend, client, and mentor Jim Britt. Jim's *Power of Letting Go* program was a huge inspiration to me during some pretty tough times. I am very happy to have his help for the second edition of this book and future *Savage* titles.

To my longtime friend Sarah Nelms, thank you. For everything.

To my partner Jennifer, I love you. This project would be all but impossible without you.

Thank you all.

Foreword

In *The Enlightened Savage*, Anthony Hernandez has written a remarkable book that combines anthropology, business, psychology, religion and common sense. This blend of disciplines helps readers learn to trust their instincts or better yet, use their instincts to best advantage.

In the world of business, instincts have taken a back seat to logic, which is a shame because instincts are almost always more trustworthy than logic. When President George W. Bush was appointing a cabinet, he invited one man, a Democrat, to join his Republican administration.

This man, Former Secretary of Defense William Cohen, wasn't sure whether or not to accept the invitation. So he turned for counsel to his five-year old son, asking, "What's more important, your heart or your mind?" His son immediately answered, "Your heart." Cohen then asked, "Why do you say that?" His son, without hesitation, explained, "Because your heart is never wrong."

Author Hernandez makes the same case in this extraordinary book. He shows us exactly why we ought to be trusting our hearts, our instincts, our primal parts. I've never thought of myself as a diseased monkey, but now I can understand why Hernandez might view me in this way.

And with that understanding come other understandings that I've gained by reading this book, insights that many readers will find to be crucial in their quest for a happy, balanced and successful life.

I know the author to be a whiz at marketing, so when I saw that this book was about evolution and genetics, I realized I was wandering into exotic terrain... yet the experience illuminated my knowledge of marketing and psychology.

Because this book is so engagingly written, it's a simple task to see the world— and yourself, in the extraordinary way that the author sees it. To some readers, his way is radical, while to others, his vision is the only view. If you're as fascinated as I am by his revelations, you're in for a good read.

It may be uncomfortable for you to learn some of these things about human beings, but it may end up with you being even more uncomfortable if you don't know them. *The Enlightened Savage* is a mind-expanding substance. It opens eyes and adds a new dimension to minds.

After you read it, you'll do at least one thing differently: You'll trust and heed your instincts and will know how to program those instincts to get what you want and deserve out of life and business. Take it from Mr. Hernandez: that's a good thing.

Jay Conrad Levinson
Puget Sound, Washington 2005

Introduction

Welcome to the second edition of *The Enlightened Savage*. I have long been fascinated by the question of why people do the things we do, in part because of the astonishing prevalence of so-called mental disorders ranging from mild depression to full-blown schizophrenia and psychosis. I found myself asking whether humanity as a species was indeed so sick and, if so, why. Could it be the many conflicting messages and expectations society heaps on us? Toxic chemicals in the environment? Our educational systems that are designed to breed conformance instead of free thought and individuality?

The more I asked these questions, the more I found myself feeling that these causative factors couldn't be the final answer, because each has its own causes. Finding the answer would require a lot of digging. Or would it? The more I dug, the more I got the nagging suspicion that the answer was staring me in the face.

After many years, it dawned on me that I was asking the wrong question. If mental and stress-related illnesses are indeed as prevalent as they appear, that would mean that humans are a pretty fragile species—too fragile to have possibly lasted through millions of years. Clearly, we humans had lost touch with our own nature. But what is that nature? This question finally got my thinking on the right track. What if modern society, with all of its problems and challenges, is not a result of our losing touch with our nature but a direct result of that nature? In other words, what if our challenges and struggles are caused, not because something is wrong with us, but because something is right?

I am not a doctor, and cannot and will not make any medical diagnosis. The brain is an organ like any other, and is therefore just as capable of illness and disorder as any other organ in our bodies. That said, the question remains: What if the vast majority of us are doing what we do and experiencing our frustrations and failures because we're perfectly healthy?

The answer is simple: If one is reading the end of a book and wants to know the back story, all one need do is flip back through the pages. I therefore decided to flip back through the pages of history and evolution. To do that, I took a closer look at our evolutionary cousins, the many animals we share this planet with. The answers I found merged into the theories that became the first edition of the *The Enlightened Savage* and that I continue to build on in this second edition.

I believe that the answers to our individual and collective ills lie in the millions of years of evolution that wired our brains and bodies to operate in certain ways. For you, that means that getting the success you want and deserve in life is as simple as learning how to use the primal instincts lying deep in your brain to your advantage. To help you do that, *The Enlightened Savage* combines human evolution with some of the ideas suggested by quantum physics and a dash or two of spirituality.

Everyone has different beliefs about creation, evolution, spirituality, and religion. I respect everyone's beliefs. I believe that evolution and creation, mind and spirit, body and soul, and free will and destiny are perfectly compatible ideas. In other words, the presence of one does not mean that the others can't be equally true. I'll be showing you how throughout this book and throughout the rest of the *Savage* series.

Here's an amazing fact: Humans have evolved over millions of years with our earliest known ancestors going back at least 4.4 million years. Modern humans have only existed for about 120,000 years. Depending how you define it, civilization has existed for less than 10,000 years. If we consider all of human evolution as fitting into a single hour, then modern humans have existed for less than 2 minutes and civilization has been around less than 10 seconds. If we only consider modern human existence as taking up the hour, then civilization has only existed for less than 5 minutes.

Either way, everything we take for granted today came along quickly—far too quickly for our bodies and minds to keep up. This means that we're still following the same instincts that have been etched into the deepest parts of our brains over millions of years. Understanding where these instincts come from and how they work is therefore crucial to becoming an enlightened savage.

As we'll see in later chapters, our brains and instincts evolved for one reason: To keep us alive. Understanding this simple truth is the key to unlocking your potential, and to living the life you want to live instead of the life you may feel forced to live.

Our instincts can either prevent us from achieving the personal and business success we want and deserve, or they can help us achieve our wildest dreams and more. Most people are slaves to their instincts. Enlightened savages understand where these instincts come from, how they work, and how to turn them to their advantage.

Most people believe that external forces govern their lives. Some interpretations of quantum physics postulate that outcomes depend heavily on the observer. This suggests that your life and its outcome depend heavily (if not totally) on you, because you create your own personal reality. Everything you see, hear, feel, taste, and smell is nothing but the result, the effect of core beliefs formed by emotional programming. This means that you have absolute power over your life. How you use that power is entirely up to you.

Spirituality and religion teach us that we are on this planet for a higher purpose. Most of our struggles come from not knowing or accepting our life's mission, or dharma. *The Enlightened Savage* is designed to show you how to become the active creator of your own destiny. I look at it this way: If there is a God or other higher power, and if each of us has an immortal soul (as I very strongly suspect), then each of us is here for a reason. It's up to us to find and fulfill that reason. And if there is no such power and this one short life is all we get? In that case, we'd better make good use of the limited time we have!

Question: Do you have all the joy, wealth, and fulfillment you want in your personal life? How about in your business life? Since you're reading this book, my guess is no on both counts. If you answered yes to one part and no to the other, then you need to reconsider your answers, because humans are holistic beings. You cannot affect one part of your life without affecting your entire life.

Second question: Have you tried other self-improvement programs? This is an important question, because whether *The Enlightened Savage* is your first or your fiftieth foray into self-improvement, I want it to work for you. I want you to get out there, live your life, and get the business and personal success you deserve—and believe me, you deserve every success!

If you have experienced other self-improvement programs without getting the results you wanted, why? There are lots of great coaches out there with lots of great information. Why didn't their material work for you? What makes *The Enlightened Savage* different? Why will this material work for you? I have two answers.

First, every program relies on a model of where success comes from and how to get it. A model is a method of explaining observed behaviors or phenomena. For example, Isaac Newton created a model of physics, and Albert Einstein created a new model based on his theory of relativity. Today, quantum physics and its offshoots are at the cutting edge of science. Each of these models attempts to explain how the universe works.

I have personally used the model contained in *The Enlightened Savage* to transform my life and the lives of other people through my coaching and consulting. That said, it is important to remember that a model is just that—a model. Other programs may not have worked for you because their models didn't resonate with you. That's perfectly OK. *The Enlightened Savage* model can work for you if you're willing to at least try it on for size. I believe it will resonate with you, and that you will let it work to its fullest potential in your life.

Second, these other programs may not have worked because you didn't take action to implement them. In this respect, *The Enlightened Savage* is no different. It can be the catalyst for sweeping change in your life or a pointless waste of time. It's up to you, because you're the only one who can decide how to use this material.

See if this sounds familiar: You decide to try a self-improvement program and dive into the book, seminar, or workshop, bound and determined to emerge a different person. This program is going to change your life! You can do it! And then you reach the end. The high fades and your old core beliefs and programming come right back. Suddenly this great life-changing program isn't working. It's too hard. You don't have time. It didn't address your specific needs. The trainer doesn't understand your needs. Besides, every program tells you something different, which can only mean that no one has all the answers. You'll get started tomorrow.

The list of possible excuses is endless. But the only common element linking everything together is you! You alone create your inner personal reality. Part of your brain desperately craves wealth, happiness, and fulfillment, and knows that your current way of doing things isn't working. The problem is that your brain is built for survival, and survival means following your core programming. Therefore, no matter how dark, dank, and rank you might believe your life to be, there you'll stay until you die. If the pain of inaction becomes greater than the pain of doing something, then you will finally make the choice to emerge.

This isn't the program's fault—and strictly speaking, it isn't your fault either, because your brain is following its designed function by faithfully executing your core programming. But while it isn't your fault, it is your responsibility. Why? Because (you're going to read this a lot) you alone create your own personal reality. Understanding this is critical to becoming an enlightened savage. As you will see, this has nothing whatsoever to do with "attraction" or "manifestation." It has everything to do with *realization*, which I define as, "the act of making real."

What makes *The Enlightened Savage* different? If you've ever listened to a child, you know that the one question they always ask is "Why?" The thirst for knowledge is a primal human drive that's right up there with the need for food and shelter. I will therefore take the time to lay a solid foundation for lasting change in your life by explaining the Enlightened Savage model in great detail. This is very important, because those who do not understand and learn from history are doomed to repeat it. After explaining the model, I'll show you easy steps to implement it by changing your core programming, and will follow up with emotional and cognitive support to nourish those changes. *The Enlightened Savage* works by addressing your core emotions, thoughts, and actions all at once.

You must be open to change in order for it to occur. This sounds simple, but you'll soon see that human brains are designed to resist change at all cost. You'll throw up all kinds of barriers to prevent change from occurring. Be warned that the most insidious and destructive barrier uses just four little words: *I already know that.* The moment that thought enters your mind, you close yourself off to growth, learning, and change. If you can keep those four little words at bay, you'll be way more open to reaping the huge rewards this program can offer you.

The choice is yours. You alone have the power to either embrace *The Enlightened Savage* or add it to your heap of discarded self-help programs. Becoming an enlightened savage means examining and possibly abandoning everything you hold sacred in order to discover what's really important. Do this, and your life will change forever. Guaranteed. The more difficult you find a certain chapter, the more benefit you will gain by sticking with it.

As I said before, this book is completely based on solid evolutionary science with a dash of quantum physics and spirituality thrown in. If you want to know even more about the evolutionary science behind this book, you should read *The Natural Savage*. That book explains how all of life breaks down into six core functions: predator avoidance, group status, food, shelter, reproduction, and death. Future *Savage* books will address each of these six areas in much more detail, starting with *The Divine Savage*, which explains what happens when we die. (Hint: The answer will absolutely amaze you!)

Sir Isaac Newton transformed the sciences by generating a comprehensive model of the physical laws governing our universe. The two main features of Newtonian physics are *determinism* and *materialism*. Determinism means that one can calculate any future or past state for a system based on the system's current state. Materialism is a world view where matter and energy hold sway. Taken together, Newton's model shows us a clockwork universe without the need for any God or other outside intervention. That idea opened the door for Darwin's theory of evolution, which shows how complex life evolved from simple forms, again with no God or outside agency required. This "classical" world view forms the scientific foundation for this book.

The advent of quantum physics and Einstein's theory of relativity revealed a very different universe. Gone are the cogs and springs of Newton's universe, replaced by a universe where not even space and time are not absolute. In fact, there is good reason to think that space and time are illusions! On the small side of the size scale, subatomic particles aren't particles at all. They are waves of probability that only "collapse" into a particle when a conscious observer looks at them. In other words, there is no such thing as objective personal reality. Personal reality is a purely subjective construct that depends almost entirely on how you choose to build it! Take this idea a little further and you can start to see how evolution itself may have caused the universe to come into being as we know it... how the first organism capable of sensing its environment caused that environment to come into being all the way back to the Big Bang of creation itself!

There is, however, a layer of shared reality that is common to everybody. How we interpret that shared reality is up to us. Personal reality thus sits both above and below shared reality. Shared reality must always obey the laws of physics, which is one big reason why the "Law of Attraction" preached by so many self-help experts is utterly useless. More on this later.

Newton's wind-up universe has been disproved. Evolution may well be the cart that came before the horse. So why am I relying on these theories for this book? That's an excellent question. The truth is that quantum physics actually contains Newtonian physics within it. Every single object obeys the laws of quantum physics but the quantum uncertainties and other "strangeness" operate at extremely tiny scales that seem to smooth themselves out the further one gets from them. It's kind of like looking at a microscopic view of an extremely rough surface that looks and feels perfectly smooth when seen the way we see things. Newtonian physics are therefore perfectly adequate for use in everyday situations because they results they give are extremely close to the quantum results—close enough to build skyscrapers, send spacecraft to distant planets, and any one of the many things we use physics for every day. Newtonian physics remain perfectly valid despite more recent discoveries and theories.

Evolution also remains perfectly valid. Look at it this way: Even if you personally created your entire evolutionary history going back to the dawn of time itself, as some theories suggest, then you still have an evolutionary legacy. That legacy is shared by all life on Earth.

But there are limits. Newtonian physics breaks down at very small (atomic) and very large (speed and distance) scales. Evolution also breaks down before birth and during/after death.

Yes, it explains how you got here via an unbroken line of ancestors going back to the first primordial cell. Yes, it explains what will happen to your body when you die and how your descendants, if any, will go on growing and changing. But evolution does not say anything about what or where you were before this lifetime, nor does it say anything about what you will become or where you will go after death. By "you," I of course mean your individual identity as a person. Materialist science (including evolution) says that your mind is generated by your brain. When your brain dies, you fade into nonexistence. That's the bad news. The good news is that materialism is almost certainly wrong. Put simply, it now seems like there is a lot more to life than the life we are all familiar with. (Read *The Divine Savage* to learn why this is true.)

With that said, evolution does a fantastic job of explaining both the powers and the limitations that exist for us during this lifetime. It explains all of the instincts, drives, and thought patterns each of us inherited over millions and billions of years and how that legacy affects every one of us at every moment of every day. Evolution is ruthless. We are all savages by definition. Learning how evolution rules our daily lives and using that knowledge to our advantage makes us enlightened savages. Becoming an enlightened savage is how you stop letting life happen to you and start making life happen for you. It is also your key to starting to unlock the quantum secrets that lurk behind the scenes.

Commit yourself to becoming an enlightened savage, and whatever future you want will be closer, easier, and more wonderful than you ever dreamed possible.

Are you ready to become an enlightened savage and to experience way more joy, fulfillment, and success in your life? Let's get started!

Anthony Hernandez

Chapter 1

Why Become an Enlightened Savage?

You are about to experience a profound life change for the better—If you want to.
Anthony Hernandez

You are about to embark on a life-changing experience that will challenge beliefs you've held as long as you've been alive with no "attraction" or "manifestation" required. (The "Law of Attraction" has nothing to do with anything in this book.) Make no mistake, this is a major undertaking that may cause some discomfort at times. Why are you doing this? To become an enlightened savage. Why become an enlightened savage? That's up to you. Everyone's reason will be different. Please take a few minutes to read the Introduction if you have not done so already because it gives you important information about how I wrote this book and why. When you are ready, let's begin finding out what makes an enlightened savage tick.

Motivation

If there is one thing that binds all of humanity, it is this simple truth: Our time on this planet is limited. If there is a next life beyond this one (as I strongly suspect there must be, then one can argue that the better we do in this life, the better prepared we'll be for what comes next. What if this life is the only ride we get on this crazy merry-go-round? Then there really is no excuse for not living each day to its fullest.

I almost always begin a new coaching relationship by asking my clients a simple, yet profound, question: In your last few moments of life (which I hope won't occur for a long time), as you look back, powerless to change anything, what do you want to see? What do you want to have done, been, experienced, contributed, and left behind? In short, what's going to be important to you then?

If something will be important to you as you prepare to leave this world, it had better be important to you now. If not? Then you had better put it out of your mind right quick, for we have no idea when the end will come. What a tragedy it will be if your last conscious thought in this lifetime is "if only!"

Death… is no more than passing from one room into another. But there's a difference for me, you know. Because in that other room I shall be able to see.
Helen Keller

If this ultimate fact of life isn't enough to motivate you to action, I don't know what will. I must therefore assume that you are reading this book because you know something needs to change in order to live the full, rich life you were born to live. You may not know how to get there and may be feeling frustrated at your lack of progress. That's OK. Just by reading this far, you've already started your journey.

Universal Goals

Each of us is a unique being with equally unique goals. That said, our goals tend to follow a very common pattern, one that will remain as you become an enlightened savage, and after.

Perfection of means and confusion of goals seem—in my opinion—to characterize our age.
Albert Einstein

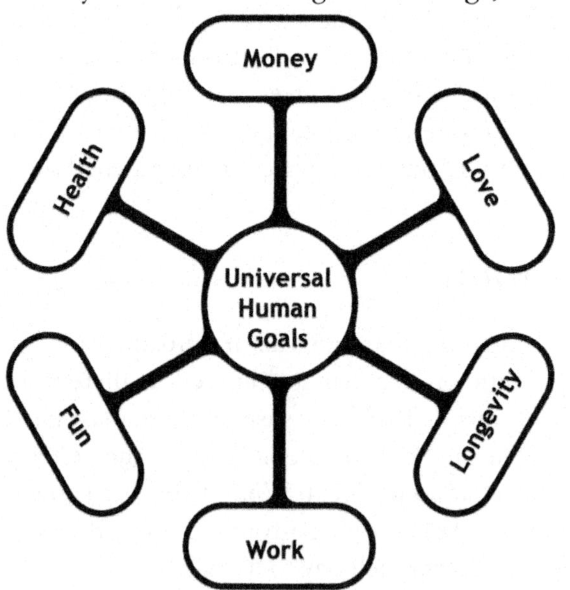

Everyone who has ever walked this earth wants the following:

Work

Work is an essential part of being alive. Your work is your identity. It tells you who you are. It's gotten so abstract. People don't work for the sake of working. They're working for a car, a new house, or a vacation. It's not the work itself that is important to them. There's such a joy in doing work well.
Kay Stepkin

We all seek to be productive in our own way and want our efforts to be a source of joy and fulfillment. We want to know that we are creating something positive and leaving a lasting legacy. The problem comes when we start defining *work* as the effort itself and losing focus on the desired results. This leads to endless struggle, which causes stress, which in turn causes a whole host of physical ailments up to and including premature death.

Embracing the idea that work is measured not in effort (struggle) but in output (results) can free us from much of our daily stress and angst. Combine this with finding and living your life's mission, and you'll be well on the path to the success you want and deserve.

Money

Superfluous wealth can buy superfluities only. Money is not required to buy one necessary of the soul.
Henry David Thoreau

Each of us seeks to have enough money so that we need never worry about it. By itself, money is nothing more than a tool, a universally accepted means of exchange that is no more good or evil than any other tool in your possession. If you own a hammer, you know that tool can be used to build homes or to destroy lives. Your pen can be used to spread love or hate. The tool itself is not the issue; our perceptions and emotions surrounding the tool are what matter.

Too many self-help programs equate wealth with a huge bank account and the many accoutrements that can come with it. An enlightened savage equates wealth with fulfilling her or his life's mission, also referred to as *dharma.* Whether you were born to live a life of financial poverty in a spiritual outpost, studying endangered species in a remote jungle, or a life of extraordinary material abundance does not matter. So long as you are fulfilling your dharma, you are wealthy. If you have lots of money but are not following your dharma, then you are truly poor.

Health

The ability to count on our bodies to respond when asked and to be free from pain is a tremendous blessing. If you or someone you know suffers from any health issues, you know all too well that health good enough to take for granted is something we all desire.

Health is a state of complete physical, mental and social well-being, and not merely the absence of disease or infirmity.
WHO Constitution

Obtaining and maintaining this sought-after health can pose a seemingly insurmountable challenge because of the many stresses we face each day. Here again, I believe that many of our real and imagined ills stem from not accepting or following our individual dharma. We'll be exploring this concept in a lot more detail later in this book.

Love

Who does not seek to build and maintain loving, supporting bonds with others? And yet we often seek this love because of some sense of emptiness or longing to satisfy some unfilled need. This leads to codependency, divorce, infidelity, and many other problems when we perceive that someone we love is no longer satisfying those needs.

Immature love says: "I love you because I need you." Mature love says: "I need you because I love you."
Erich Fromm

How wonderful it would be to seek love from a place of wholeness, where the love itself is the need instead of an outgrowth of trying to satisfy other needs. This very subtle difference is not always easy to discern, yet the results are tremendous.

Fun

Everyone loves having fun. Enlightened savages seek fun in everything they do and are, from the very essence of everyday life. They don't chase fun and never segregate their lives into "drudgery" and "fun." Every day, every waking moment is fun to the enlightened savage.

The more complex the mind, the more pressing the need for simple play.
Leonard Nimoy
(as Mr. Spock)

Longevity

If nothing else, a long life postpones the inevitable. I assume that you are seeking to live as long as possible, but why? Is your only goal to avoid death as long as possible? Or do you seek, as I hope, to live long enough to gain the wisdom to truly appreciate all of your accomplishments?

Life well spent is long.
Leonardo Da Vinci

I've lived just long enough to know that the path to appreciation is built on satisfying work. Remember that science defines work not by the effort but by the results.

Getting There

One's destination is never a place, but rather a new way of looking at things.
Henry MIller

The specific ways in which these universal goals reveal themselves in life are unimportant. What is important is realizing that our time here is limited, and that we owe it to ourselves and everyone around us to make the most of it. If we don't, who will?

Just having these goals is not enough, because we all have these goals. What sets enlightened savages apart is that they consistently look for ways to achieve each of these six goals in everything they do every day, and then act.

Stay Strong!

In the next few chapters, I'm going to explain how evolution has shaped your entire life. I'm going to explain the source of your troubles (and joys) and am going to show you how your built-in biological limitations are keeping you from living the life you want. This may be difficult material to read, and you may find yourself feeling powerless and helpless in the face of millions of years of evolved instincts. You may find yourself feeling sad and may even wonder what hope or chance you have of making things better.

Still, it is imperative that you read this entire book. You must know and confront your limitations before you can possibly use your strengths. Therefore, the bad news is that I do have to rub your nose in your own savage brain and primal mental programming. The good news is that you can use these primitive aspects of your mind and brain to great advantage. Stick with me though the next few chapters and I'll show you how knowing all of what I am to explain can set you free. You have great limitations, this much is unavoidable. But within those limitations lies strength and power on a scale you may never have imagined. That's a promise. Stay strong!

Questions

Why do you want to become an enlightened savage? We're going to delve into your goals and life's mission in a lot more detail later. For now, though, it's important to clarify why you bought this book and what your expectations are for the journey you are about to take.

Write down your answers to the following questions:

- **Why did you buy this book?**
- **Why will this book work when other programs have failed?**
- **If you could do or be anything you want, what would that be?**
- **What do you want to see when you look back at the end of your life?**
- **Do you consider money a tool, or do you attach emotions or other baggage to it? If so, what are they?**
- **Do you suffer from any diseases or ailments? If so, are you receiving proper care from a qualified professional?**
- **Are you close to one or more people? If you are seeking more close bonds with others, what, if anything, is preventing you from obtaining them?**
- **Do you have enough fun in your life? Would you like to have fun no matter what you're doing?**
- **What specifically do you hope to accomplish using *The Enlightened Savage*?**

Chapter 2

Rise of the Prey Savages

Jung believes that the human mind contains archaic remnants, residues of the long history and evolution of mankind. In the unconscious, primordial "universally human images" lie dormant. Those primordial images are the most ancient, universal and "deep" thoughts of mankind. Since they embody feelings as much as thought, they are properly "thought feelings."
Patrick Mullahy

To understand how and why humans do the things we do, we must begin by examining our evolutionary roots. Knowing where we come from helps us see where we are and provides hints about how our brains work. This is important, because every one of your thoughts, feelings, beliefs, and memories is completely dependent on your brain's capabilities and limitations. In other words, your entire life and personal reality depends on what your brain can and cannot do. Learning how our brains work is the key to unlocking our true potential and becoming enlightened savages.

One Common Ancestor

Tracing the paths of human and monkey evolution back through time leads us to a common ancestor species, the proverbial missing link. Something happened to this one animal that caused humans and monkeys to split into two separate but very similar evolutionary lines (see the bottom of the following diagram). We know this because any comparison between man and monkey yields far more similarities than differences.

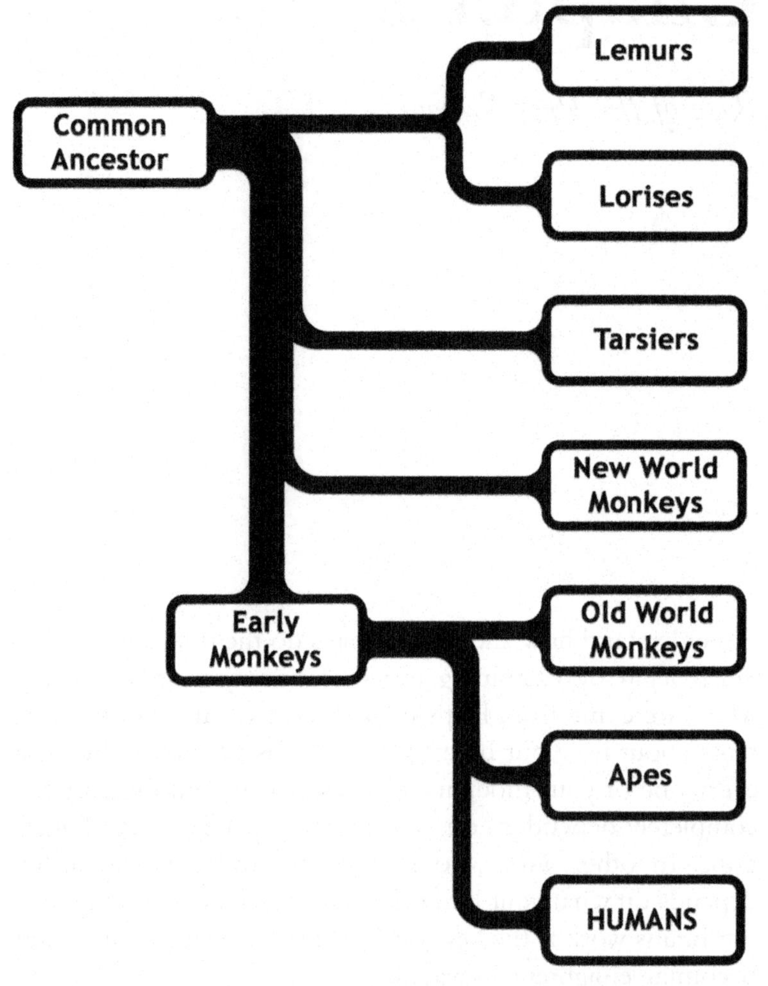

[It] is not a monkey, not an ape and not a human, but it's a common ancestor of them all.
Russell L. Ciochon

In fact, humans share between 95 and 99% of our DNA with chimpanzees, depending on which theory you subscribe to. If you compare monkeys to monkeys, the similarities are even greater. It is therefore very helpful to examine monkeys when seeking answers about human behavior. Let's take a look at our primate cousins.

The Human Ape

A monkey is a monkey is a monkey. Some live in trees while others live on grasslands, and many live in varying climates. This is possible because nature perfectly adapted each species to thrive in its own unique habitat. So what about humans who evolved from the common ancestor, spread over the globe, and adapted to our own local habitats?

Being abroad makes you conscious of the whole imitative side of human behavior. The ape in man.
Mary McCarthy

What I'm about to say may sound controversial, but please bear with me: I believe that the different species of monkeys are roughly comparable to the different human races—and what a beautiful example of Nature's adaptability! Caucasians evolved in cold areas with weak sun and relied on clothing for warmth. Our skin may be pale because sunburn wasn't a big hazard. Asian eyes may have evolved to protect against flying sand and debris in the cold, windy Mongolian deserts. Races that evolved in warmer climes may have evolved varying degrees of skin pigmentation to ward off sunburn.

Like monkey species, race may be a physical result of evolution accounting for and adapting to different environments. With that said, there is sameness beneath the obvious differences: All human beings of all races share the same intellect, the same spirit, the same heart, and the same capacities for joy, love, happiness, and success. I believe this is one of millions of beautiful examples of Mother Nature at work.

Diseased Monkeys

Nevertheless, the difference in mind between man and the higher animals, great as it is, is one of degree and not of kind.
Charles Darwin

Species and race aside, there are too many physical similarities between humans and monkeys to list. But for one key difference, humans might never have evolved as a distinct species. What's this major difference? Ah, now that's the question.

A scientific discovery made in early 2004 gives us a tantalizing clue: It seems that the single largest difference between humans and apes may lie in a genetic mutation that occurred about 2.4 million years ago. This mutation caused human jaws to weaken, with two effects: First, an animal with a weakened jaw can't bite as hard, meaning that it's just lost a critical defense against predators. We'll revisit this very important concept in the next section. Second, a gentler bite places less stress on the skull.

Here's where it gets really interesting, because reducing stresses on the skull allows both the brain case and the living computer within to expand in both size and capabilities.

If this theory is true—if one of the single biggest genetic differences between humans and monkeys is indeed the gene that's mutated in humans—then human evolution began with a genetic disorder. In other words, we humans may arguably be diseased monkeys. That's a rather humbling thought. It's also an empowering thought, because we can turn to our evolutionary cousins for insights into what makes us tick and how we can use the primal instincts hard-wired into our brains to our benefit.

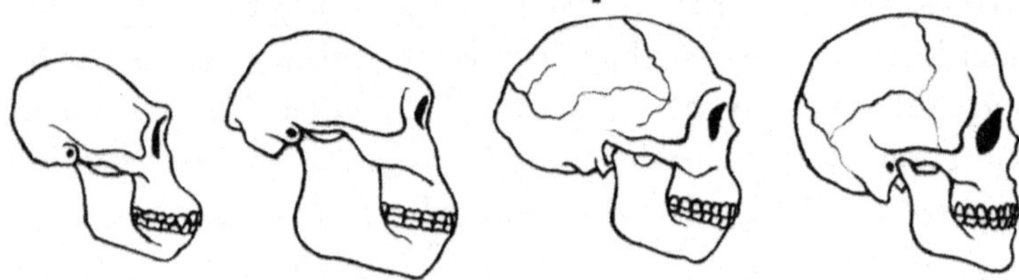

Swapping Brawn for Brains

> *I was taught that the human brain was the crowning glory of evolution so far, but I think it's a very poor scheme for survival.*
> Kurt Vonnegut, Jr.

The transformations described above aren't limited to the jaw and brain: One look at our evolutionary ancestors reveals that humans traded raw physical strength for unmatched mental ability pretty much across the board.

Monkeys are far stronger that the average human on a pound-for-pound basis. Mentally, the exact opposite is true. As humans lost the ability to survive with our brawn, we gained the ability to survive using our wits. Thus humans evolved, carrying their evolutionary instincts with them. Many people use the term "savage" to describe our earliest ancestors, who presumably lacked the social graces we take for granted today.

Today, the instincts that evolved for our survival can actually prevent us from achieving the success we want and deserve—not because they're bad or malfunctioning, but because they're intact and working as intended. I'll cover this in a lot more detail later. Meanwhile, I'm using the term *enlightened savage* to describe those of us who learn to use the instincts we evolved with to achieve the personal and business success we seek.

Built-in Limits

> *The Intentionality of the mind not only creates the possibility of meaning, but limits its forms.*
> John Searle

Let's take a closer look deep inside our heads. If you've ever used a computer, you know that the hardware is the actual machinery, while the software is simply electromagnetic information stored and processed by the hardware. You also know that the software depends on the hardware in order to run. If your computer malfunctions, you can't use your software. Software also depends on the hardware for its capabilities. For example, if you have an audio program but no speakers, then you won't hear any sound, no matter how much you adjust the program's settings. How does this apply to you?

Cause and effect are two sides of one fact.
Ralph Waldo Emerson

A human brain is like an amazing piece of computer hardware that contains and executes software programs that together form every one of your thoughts, feelings, beliefs, and memories. Please don't get me wrong: I'm certainly not suggesting that people are nothing but glorified computers. I see it this way: If we are nothing but biological creatures whose brains are the source of our consciousness (which I strongly doubt), then we are constrained by our brain's abilities and limitations. If we possess a soul or spirit, and our brains are merely conduits for consciousness, then we remain constrained by the same limits, in the same way that one cannot obtain a high-definition TV picture from a 1950s TV set, no matter how good or strong the signal might be.

In either case, exploring the root cause behind your life, with the trials and frustrations that prompted you to buy this book, forces us to start at the beginning: Your hardware. Your brain.

Humans as Prey

Humans are such easy prey.
Stuart Gordon

I assert that humans are prey animals. I do this for two reasons: First, humans and monkeys evolved from the same common ancestor, meaning that we share common evolutionary roots. Second, monkeys are routinely killed and eaten by predator animals.

Can man and ape be predators? Certainly. Humans have hunted animals for thousands of years. Some monkeys also eat meat, most notably chimpanzees. However, until very recently in our evolution, being killed and eaten by saber-toothed tigers and other creatures was a very real possibility, a daily threat.

For most people alive today, being killed and eaten is nothing more than a fairy tale. However, wild animals still do hunt and kill humans in remote corners of the world. Our monkey cousins face the threat of predation at every single moment of every single day.

This is an extremely important concept to understand because modern civilization has only existed for a few thousand years, while our predators have hunted us for millions of years. Our brains literally have not had enough time to rewire themselves to reflect our modern shared reality. This is why I assert that our prey instincts are alive and well in the core of monkey and human brains and personalities, and that these instincts are the root cause behind why we do what we do and why we behave the way we behave.

Sound crazy? I thought so too until I took a closer look.

What's Eating Us?

A monkey's chief predators are pythons in the trees, big cats such as lions and tigers on the ground, and birds of prey in the air. Panama's harpy eagle is the world's largest *raptor* (bird of prey). While most eagles are designed to soar through the sky, the harpy eagle is purpose-built for high-speed flight through the forest canopy. Its extreme size and maneuverability lets it dodge branches and carry very heavy loads. Its chief diet? Monkeys.

Feet are considered a delicacy among certain animals, you know... In fact, there are certain man-eating animals who will eat only the feet, leave everything else, will not touch one other thing.
Blake Edwards

People in military circles use the term *combined arms* to describe an attack that occurs from all sides at once, leaving the enemy no place to run or hide. With lions on the ground, snakes in the trees, and eagles overhead, no monkey can ever be truly safe, because there is always an animal waiting to kill and eat it.

This constant peril from all sides is a perfect example of combined arms where the hapless monkey is left bereft of refuges. As creatures who evolved from and alongside monkeys, humans share that perilous legacy.

Getting Warmer

Speaking of humans, what is humankind's single greatest technological leap? Here are three hints:

O for a muse of fire, that would ascend the brightest heaven of invention.
William Shakespeare

- This technological leap involved harnessing a natural phenomenon.
- It occurred about 400,000 years ago.
- This same leap is responsible for just about every aspect of modern life.

Go on, take a guess. Did you say fire? If so, you're absolutely right.

Take a look around at just about any object you want. Fire in some form plays a direct or indirect role in just about everything we take for granted today, from our food to our products. Therefore, harnessing fire is arguably the single greatest technological leap that humans have ever made.

Let's think about the many ways in which fire touches our lives. What's more romantic than sitting by a fire? What's a great camping trip without a campfire? What's great food if it's not served piping hot—or even better, barbecued? What about our energy sources? With rare exception they involve fire in some form. And our weapons? You see my point.

Fire is an incredible force that's at once life-giving and life-taking. A forest fire causes death and destruction, but also thins the forest and allows new life to grow. Some pinecones even need fire in order to release their seeds and reinvigorate the forest. In fact, over a century of fire prevention efforts have left many forests choking and dying under massive overgrowth, which makes them susceptible to truly devastating fires that incinerate all life.

Fire breathes. No oxygen, no fire. Some people describe fire as an animal. Since fire breathes, eats, and can give or take life, that description sounds pretty accurate to me.

The Ultimate Predator

Children today laugh at fathers who tell them about dragons.
Karl Krause

What do you get when you cross a snake's body and scales, a big cat's legs and fangs, a bird of prey's wings and talons, and fire? Here's a hint: This mythical creature appears in just about every ancient, medieval, and contemporary culture on Earth. In all cases, this creature shares many universal characteristics.

This terrible beast is a dragon. Dragons and dragon-like creatures are the one family of mythical creatures that transcends time and culture. No group of people will ever share the same beliefs, superstitions, religions, etc. But they will all recognize a dragon.

> *Do you want to have an easy life? Then always stay with the herd and lose yourself in the herd.*
> Friedrich Nietzsche

These facts have me convinced that humans are prey animals with images of our predators forever etched into our primordial memories and instincts. How about you? You can choose to buy into this model or not. The beauty is that you don't necessarily have to accept the model in order to reap its benefits.

In computer terms, building a specific program into the machine's actual, physical circuitry is called hard-wiring. I believe that our prey instincts and images are literally hardwired into our brain's circuits.

How else can one explain the presence of the same mythical creature that combines our cousins' three primary predators with our single greatest technological advance across civilizations that had no way to communicate with each other?

We recognize cats, snakes, and eagles so easily because they are our evolutionary predators. We also have intense primal instincts and memories around fire that have been accumulating for almost half a million years. Why else do we gravitate toward it so strongly? Why are homes with fireplaces so much more valuable than homes without them? Why are campfires such an essential part of the camping experience?

As a side question, if dragons represent the sum total of our greatest achievement and our three evolutionary predators, what about the dragon slayer? Now there's a superhero! Predictably enough, mythology reserves some of its highest praise for that rare breed of human who can stare down and kill that which represents the sum total of that which would literally consume us.

Further Evidence

Do you not see how necessary a world of pains and troubles is to school an intelligence and make it a soul?.
John Keats

An article published in January of 2006 quotes Lee Berger, a paleo-anthropologist at Johannesburg's University of Witwatersrand, as saying that his research indicates that our evolutionary ancestors were hunted by birds. Studies of thousands of monkey skulls revealed injury patterns such as holes and jagged cuts behind the eye sockets that were caused by raptor attacks.

These studies allowed Berger to determine that the cuts and holes in the two-million-year old *australopethicus africanus* skull known as the Taung child skull are consistent with being killed by a predatory bird such as a large eagle. Berger concluded that our ancestors had to survive being hunted from both the ground and air. According to Berger, discoveries like this are "key to understanding why we humans today view the world the way we do."

I couldn't agree more.

Welcome to the Herd

Competition has been shown to be useful up to a certain point and no further, but cooperation, which is the thing we must strive for today, begins where competition leaves off.
Franklin D. Roosevelt

Apropos of slaying dragons, what about monkey versus predator? Pit any monkey against a cat, python, or eagle, and my money is on the predator. Your basic monkey just isn't strong enough to fight off a predator. It also lacks the ability to use weapons. A lone monkey is therefore in a pretty dire position.

But wait... If one monkey can't prevail, maybe a group of monkeys can. Well guess what: monkeys are social animals.

There's strength in numbers, strength for defending the troop, securing feeding grounds, and keeping other troops away from the home turf. Groups make lots of evolutionary sense: A lone monkey has a 100% chance of being attacked by a passing hungry predator. A troop of 25 monkeys drops the odds to 4% that any one will be attacked, all else being equal. I don't know about you, but I'll take 4% over 100% any day.

The odds get even better because a group of monkeys working together might be able to fight off the predator and save the entire group. Yes, there is certainly strength in numbers, especially when one enemy is stronger than one defender.

There's just one catch to this: In order to be effective, a group must work as a team with common goals and purposes.

Follow the Leader

Left to its own devices, Nature is pretty competitive. It tries to pick the best and the brightest to reproduce, theoretically improving the species with each generation. Any manager will tell you that a group of individuals running around doing their own thing can't possibly be very effective at fending off predators or at giving the strongest monkeys their chance to shine.

I am a leader by default, only because nature does not allow a vacuum.
Bishop Desmond Tutu

Solving this problem requires establishing and maintaining leadership and a pecking order. Theoretically speaking, the strongest, smartest, and fastest monkeys have the best chances of beating everyone else to the punch and getting the rewards. They become the leaders who provide the all-important cohesion and direction.

The same thing is true for humans. We are social animals who do better in groups, to the point that we develop physical and mental disorders if left alone too long. Every group, from your circle of friends to the PTA, Scout troop, corporation, and military, has people in leadership roles directing the actions of followers under them. The degree of control a leader can exert on followers is downright eerie. Behavioral experiments have demonstrated that groups of people can be directed to actions that no individual would condone on her or his own.

Instinct leads, logic does but follow.
William James

Want some examples? How about the college experiments where students were induced to give others electrical shocks? The shocks themselves were fake but the participants did not know this. Thinking they were administering the real thing, they kept delivering shocks even when the recipients howled in mock pain and the instruments indicated that the shocks were strong enough to cause harm. Another experiment had college students exhibiting Nazi-like behavior against their peers. How about The People's Temple cult, whose leader, Jim Jones, inspired hundreds to drink cyanide-laced Kool-Aid in Jonestown? The list goes on and on.

Here's a personal example: I have some EMT (Emergency Medical Technician) training. I also frequently drive our nation's freeways. Every so often, I come across serious car accidents before emergency personnel have arrived at the scene. Without exception, there are always people milling about aimlessly. I get these folks to help me by calling 911, warning oncoming traffic, and other things. They do it without question! People like, perhaps even need, to be told what to do, and will gladly follow a leader.

Act like a leader, and chances are that others will follow you, sometimes with a shocking degree of willingness.

The Importance of Fitting In

Who knows what true loneliness is—not the conventional word, but the naked terror?
Joseph Conrad

The science of rank theory postulates that depression is an evolved response to a loss of rank and subsequent belief that one is a loser. Depression eases the sting of losing and helps people cope with defeat. This prevents the loser from suffering further injury, while also preserving the stability and efficiency of the group by reducing aggressiveness and establishing a pecking order for food and other necessities.

Look at children and adolescents. Fitting into and being part of the group are incredibly important. Everyone seeks and jockeys for position, popularity, and rank. Those who fail tend to be depressed. Those unlucky enough to be expelled from the group suffer a major disaster: Deep in our subconscious, our prey brains know that safety comes from numbers.

If the foregoing is accurate, then our status as social, hierarchical creatures is a direct evolutionary descendant of behavior evolved by our evolutionary ancestors because of the need to counter predators. As such, I assert that this thing we call civilization is nothing more than prey animal instincts in action under the ever-present shadow of our inner dragon.

Prey Behavior

There exists a fundamental difference between the way prey and predator animals behave. How do most prey animals respond when thrust into a new situation? Chances are they'll move tentatively and seek to retreat or cower when confronted or startled. Startle a mouse and odds are excellent that it will run for cover. What about predators? Have you ever played with a kitten? Startle it or do something unexpected, and its first response is to hiss or even swipe its claws. Predators stand and fight; prey runs.

It is asserted that the dogs keep running when they drink at the Nile, for fear of becoming a prey to the voracity of the crocodile.
Pliny the Elder

Take a close look the next time you watch a nature show. Pay careful attention to the prey animals, who are always vigilant, always on guard, always nervous, always ready to cut and run. In groups, some are always pulling guard duty while others go about their business. As for the predators, the lions and tigers are relaxing fat and happy in plain view of everybody without a care in the world because nobody's going to mess with them.

I believe that humans are prey animals whose every action stems from the drive to keep from getting killed and eaten. Our brains contain imprints of our chief predators, and those images appear throughout our myths and religions. Our tribes and nations evolved from the need to band together against predators.

Putting It All Together

It would indeed be the ultimate tragedy if the history of the human race proved to be nothing more noble than the story of an ape playing with a box of matches on a petrol dump.
David Ormsby-Gore

The concept that humans are prey animals who evolved certain instincts to cope with that fact is the key to understanding why we don't have the level of joy and success in our lives that we want and deserve. It is also the key to learning how to overcome our failures and become enlightened savages.

I realize that the idea of humans as prey animals who evolved from monkeys might not sit too well with you. That's perfectly understandable. Please bear with me if this idea is making you uncomfortable. In the coming chapters, I will demonstrate how this apparent weakness is actually the source of amazing power that has the capacity to completely change our lives for the better.

Please do not interpret anything I am saying to mean that humans are mindless drones, nor am I denying the existence of spirit or soul. Quite the contrary. Evolution and creation are not mutually exclusive ideas where one means not the other. I'll repeat myself because this is an extremely critical point: The presence of evolution does not mean the lack of Creation. And the fact that our brains are hard-wired with predator images in no way implies a lack of spirit or soul. Remember: What I am proposing in this book is a model. If the model helps you improve your life, then it's a good model regardless of its limitations.

It may be tough for you to realize that you are a prey animal at heart, and there are more startling revelations ahead. Stay with me, and I think you'll like where I'm taking you.

Questions

How do you feel about the idea of humans evolving from prey animals? Do you find yourself thinking "A-ha, now I get it!" or do you reject the concept?

Write down your answers to the following questions:

- **How do you feel about the notion that human are prey animals?**
- **Does the idea of looking to animals for insights into human behavior resonate with you? Why or why not?**
- **Do you believe that evolution and creation are compatible or not? Why?**
- **Assuming you agree with the concepts presented in this chapter, do you have a sense of starting to understand why you are the way you are? How so?**
- **Assuming you disagree with the concepts presented in this chapter, do you believe that the Enlightened Savage model can still help you overcome your barriers to success? Why or why not? If so, how?**

Chapter 3

The Software

Our brains are the computer hardware that contains images of our evolutionary predators and the drive to survive. In this chapter, I'll expand on this concept by showing you how our emotional programming generates the core beliefs that fit into this model and how all of this affects our daily lives.

Staying Alive

Don't criticize a snake for not having a horn; for all you know it may someday become a dragon.
Chinese proverb

What do most children fear? Monsters! Monsters lying in wait at night under the bed or in their closets to jump out and "get" them. Things that go "bump" in the night. In evolutionary terms, they fear predators waiting to invade their refuges and attack them in their sleep. Why is the human brain wired to be frightened of monsters and noises? Because our evolutionary ancestors learned millions of years ago that the thing going bump in the night is a hungry snake coming into its nest. Staring into an attacking python's eyes sure fits my definition of monstrous.

Survival Habits

Humans of all ages establish and stick to routines, no matter how bad they are. Routines are well-known, comfortable, and safe. Why? Here's an example: If you're a prey animal and your predator is out at dusk, your pattern might be to sleep at night, go out and about during the day, and be back in your burrow well before sundown. If your predator comes out at high noon, you might do well to be nocturnal.

Prey habits revolve around avoiding predators. Humans are prey animals whose number one goal is to avoid being killed and eaten by our predators. Our habit-forming tendencies exist to serve this very purpose. Once formed, you will stick to your routines no matter how inconvenient, uncomfortable, or awful they are.

Why? Because straying from that routine is the metaphorical equivalent of a gopher coming out of its burrow at high noon when it knows that hungry predators are going to be circling overhead at that exact time. In other words, deviating from an established routine violates the prey instinct. Your brain will therefore do absolutely everything in its power to keep that from happening. Have you ever heard the saying that people refuse change until the pain of not changing exceeds the pain of changing? Here is an example of that saying in action:

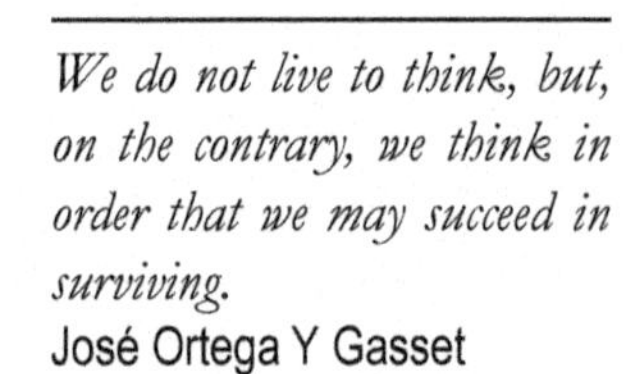
We do not live to think, but, on the contrary, we think in order that we may succeed in surviving.
José Ortega Y Gasset

A gopher whose burrow is flooding faces possible death by predation by coming up for air and certain death by drowning if he remains underground. In this exceptional situation, the pain of following his normal routine by staying put becomes greater than the pain of violating that routine by leaving the burrow. The gopher emerges to escape the water and for no other reason.

The Comfort Zone

The closing of a door can bring blessed privacy and comfort—the opening, terror.
Andy Rooney

Our routines form our comfort zone. We speak of doing something that we're not accustomed to doing as "leaving the comfort zone." Think about a time you left your own comfort zone. How did you feel? Scared? Nervous? Insecure? Naked? Ready to bolt at a moment's notice? This is a perfect example of our prey instinct in action, telling us to scurry the heck back inside our safe burrows. The farther we stray from our comfort zone, the stronger the urge to run back.

Why? Because as prey animals, our top priority is to avoid getting killed and eaten. Our brain is therefore wired to learn and follow habits designed to keep us alive. It does this by learning how our predators operate and devising patterns to counter them. This drive is so fundamental that it is literally built into our physical circuitry, our hardware. What about food, shelter, and reproduction? All of these drives are a distant second to the drive to survive. After all, it's kind of hard to be fruitful and multiply if you're inside a predator's stomach, isn't it?

I assert that human "civilization" is nothing but a very thin layer atop a slightly thicker layer of cognitive and analytical powers that in turn reside above our subconscious primal instincts. If this is the case, then saying that our thin layer of civilization can hope to compete with millions of years of evolution is rather like saying that one can stop a loaded freight train with a dirty look. One can't. Civilization works in accordance with our subconscious primal instincts and core beliefs, not the other way around.

The more comparisons one draws between humans and apes, between human behavior and prey animal behavior, the more apparent this conclusion becomes.

Act First, Ask Later

We often disguise our reflexes as deliberate actions.
Mason Cooley

Studies of wild chimpanzees in Africa demonstrate some uncanny similarities between chimp and human cultures. Books on animal behavior, such as *When Elephants Weep* by Jeffrey Masson, show an astounding degree of human-like behavior among animals.

Question: Are all these different species really exhibiting human-like behavior? Or are humans exhibiting animal-like behavior? I believe that the latter explanation is the correct one given everything I've covered so far. If this is true, then humans behave far more like animals than any of us wants to admit, because our brains have not had time to evolve in response to modern civilization. In short, we're animals—prey animals. But you knew that already.

A prey animal must learn and adapt to its predators' habits as quickly as possible without any indecision at all. Indecision leads to hesitation. That hesitation gives your predator the extra split-second to jump on you, resulting in—you guessed it—getting killed and eaten!

In the case of social animals such as monkeys and humans, the entire group suffers when a predator gets lucky, because each loss both increases the odds of everyone else being killed and eaten and decreases the group's strength and effectiveness. If 25 monkeys become 24, then each monkey's chances of being the one selected to get killed and eaten rise, threatening the entire group. The more monkeys perish, the greater the risk for those who remain.

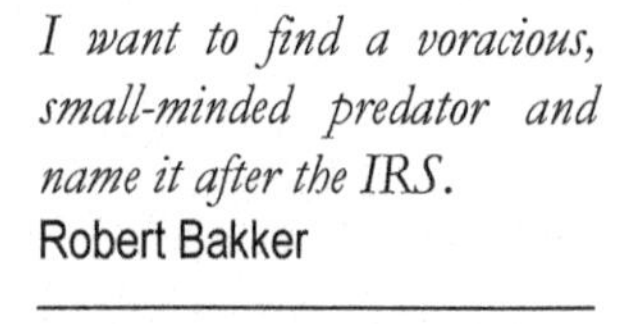
I want to find a voracious, small-minded predator and name it after the IRS.
Robert Bakker

Take a good look at monkeys, birds, and/or any species where one or both parents raise the young. In every case, the juveniles stick around to learn the survival skills crucial for staying alive. Cats learn to hunt, geese to swim, and so on. Prey animals learn when they can come out, where to go, how to avoid predators, and when to retreat to the relative safety of their burrows, nests, or trees. Survival depends on absorbing this information and mastering the skills without question. Remember that questions lead to indecision, which leads to hesitation, which gives the predator the advantage.

Once the juvenile is mature, it leaves its parents and strikes out on its own, eventually passing on its knowledge to its own children. Thus the endless cycle of life continues.

Programming the Human Animal

The child who takes his first steps and finds himself walking alone, this moment must bring the first sharp sense of the uniqueness and separateness of his body and his person, the discovery of the solitary self.
Selma H. Fraiberg

Like other animals, humans form most of our beliefs about how the world works as young children. Any child development specialist will tell you that no child has the rational, analytical, or cognitive skills of an adult. In fact, children don't even develop a sense of individuality until about age two, what we call the terrible twos. Until then, they have no concept of themselves as being separate from their environments and are therefore one with the world—the perfect state in which to receive emotionally-based survival programming, since the brain has not yet developed any barriers to receiving information. The transition from this state to an awareness of self as an individual separate from one's environment takes years to complete.

My son Logan experienced this transition in early 2004. He's always known and responded to his name. Until age two, his response to "Where's Logan?" was a blank stare. Then one day he responded by pointing to himself. That was when I knew that the separation had occurred, and when Logan understood his individuality and the concept of "I".

Loading the Data

How a child is taught affects his image of himself, which in turn, influences what he will dare and care to try to learn. The interdependence of the two is inescapable.
Barbara Biber

If you are one with your surroundings and lack the cognitive ability to analyze or question them, then it should follow that your environment and everything in it equals truth, as far as you're concerned. This complete openness, with its lack of differentiation of self and environment, allows children to absorb and act on fundamental emotional lessons about how the world works and what behavior is appropriate. Ever heard the saying "children are like sponges?" Here's where it comes from.

A child's earliest years are the foundation upon which all future learning is built. All of this future learning (including *The Enlightened Savage*) occurs after—or on top of—this very early imprinting. The inability to separate experience from self means that children absorb information in the form of emotional programming that in turn leads to the development of core beliefs. These core beliefs literally become part of the child, and direct her or his behavior without any conscious thought.

Running the Program

Our brains are hard-wired to interpret our core beliefs as programming containing the critical survival instructions that tell us who our predators are and how to avoid getting killed and eaten by them. This programming is executed deep within our brains where we are only rarely aware of its existence. Our conscious mind represents only a very tiny fraction of our brain's total processing.

Beliefs constitute the basic stratum, that which lies deepest, in the architecture of our life. By them we live, and by the same token we rarely think of them.
José Ortega Y Gasset

The comfort zone I talked about earlier is, therefore, nothing more or less than a set of behavioral boundaries imposed by the core beliefs that form when we are very young—too young to do anything about them. Too young, in fact, to know what's happening to us.

These core beliefs color a person's entire personal reality from birth until death, or until s/he decides to change them, whichever comes first. Since you're reading this book, my sincere hope is that the latter will occur first.

Core Beliefs

Let me give you a quick example of how beliefs work by comparing beliefs to sunglasses. If you put on a pair of sunglasses, they will alter incoming light rays before they hit your eye. Has the world changed? Of course not; only your perception of the world has changed. Put sunglasses on and you'll be very aware of their effects for the first few minutes.

Every man feels that perception gives him an invincible belief of the existence of that which he perceives; and that this belief is not the effect of reasoning, but the immediate consequence of perception.
Thomas Reid

Then what happens? As you go about your day, you'll forget all about the sunglasses. But they're still there, and they're still coloring your perception. They'll keep doing this until you reach up and take them off. The same thing is true of your core beliefs.

The RAS

In essence, the Reticular Activator is filtering your thinking. The incredible thing about it is that it can be programmed by your own objectives.
Wanda Loskot

How do core beliefs affect us? Our brains contain a small area called the Reticular Activation System or RAS. The RAS is a group of cells that filters incoming information as being either important or not important. A quick example of how the RAS works is walking through an airport terminal completely tuned out to the many sounds around you until you hear your flight called and prick up your ears.

The RAS determines which portions of the mass of information around you will make it into your conscious awareness. This is an extremely tiny percentage: According to one source, your brain processes some 400 billion bits of information per second. That's 400,000,000,000. Of these, you are aware of only about 2,000, according to the same source. That means your brain processes 200 million bits of information for every one you're aware of. How's that for effective filtration? Note that the numbers themselves aren't too important. What is important is understanding that we are only aware of a tiny percentage of all that is happening inside our brains.

The RAS is critical for survival because it learns how to separate things that might get you killed and eaten from all the distractions surrounding you. Imagine sitting in a crowded restaurant. If you're focusing on your conversation with your date, you'll hear your conversation clearly, with the other hubbub being nothing but vague noise. Why? Your reticular activator system.

In general, your reticular activator bases its filtration activities on your core beliefs, which result from emotional programming loaded into your brain at a very young age. Barring any changes, all of the beliefs, knowledge, thoughts, and feelings that you will ever form throughout your life will be 100% consistent with that initial programming. As a prey animal, any failure to follow your core beliefs risks getting you killed and eaten, and your brain's number one mission is to keep that from happening.

How Beliefs Work

The essence of belief is the establishment of a habit; and different beliefs are distinguished by the different modes of action to which they give rise.
Charles Sanders Pierce

The dictionary defines "belief" as the mental acceptance of and conviction in the truth, actuality, or validity of something. Therefore, if you believe something, it is true as far as you're concerned. In other words, the only reason two plus two equals four is because you believe it does.

If you happen to believe that two plus two equals five, then there is an excellent chance that no amount of argument or demonstration is going to change that. Belief equals truth because your beliefs are the mental sunglasses that are coloring your raw sensory input before you ever perceive it. The personal reality that you perceive (everything that you feel, touch, taste, smell, and hear) is nothing more or less than the result of your perceptions. Your reticular activator picks out what's important based on your beliefs, which filter and color every piece of raw data you receive. Then, and only then, do you perceive the already-filtered data.

If your personal reality is based on your core beliefs, then the universe you inhabit is nothing more than the effect caused by your core beliefs that transform raw stimuli into your reality. Have you ever heard the statement "I'll believe it when I see it?" This statement is backward: You see it because you believe it. If you don't believe it, you don't see it. This is far more than mere semantics; quantum physics may imply that each of us creates our own personal reality from the Big Bang to this very moment—a theory called *biocentrism*. *The Divine Savage* explores this theory in much more detail.

To summarize, you formed your core beliefs from the emotional programming you received between the time you were conceived and about age two. These beliefs are shaping your personal reality and everything in it. Your universe is the effect and you are the cause.

We arrived at this conclusion by building logical bridges, but how do we know we're right? The answer lies in quantum physics, which indicates that the observer influences the outcome of any experiment. No observer, no outcome. Thus, if a tree falls in the forest and no one is around to observe it, then it makes no sound. If you find a downed tree, you can infer that it was standing and either fell or was chopped down. But unless you actually witnessed it falling, there was no event. As far as you are concerned, the tree has always been on the ground.

There's No Escape

There is no exit from the circle of one's beliefs.
Keith Lehrer

Initial emotional programming generates core beliefs that color every bit of information your brain receives, and direct your every thought and action, to create results that are 100% compatible with the original programming. These beliefs create the perfect self-fulfilling prophesy that becomes our comfort zone, our way of avoiding the predators trying to kill and eat us.

Computer software is completely dependent on the hardware. Your brain is the hardware and your core beliefs are the deepest layer of software. As a computer designed to prevent you from getting killed and eaten, it interprets your core beliefs as the instructions it needs to carrying out its survival mission.

As the song says, you can't win, you can't break even, and you can't get out of the game. That's the bad news. The good news is that you can direct the game toward any outcome you want.

Questions

Are you starting to see how your earliest life experiences shape who you are today and all of the events in your life?

Write down your answers to the following questions:

- **Do you believe that you create your own personal reality? Why or why not?**

- **Do you understand that your personal reality is limited by shared reality and the laws of physics?**

- **Do you understand that your comfort zone exists as a safety mechanism? How does this understanding change your perspectives on why you may not be living the life you want and deserve to be living?**

- **What if you could do and be whatever you want and adapt your comfort zone to suit your wants and needs instead of the other way around? What would your life look like?**

- **Can you identify any specific events that influenced who you are today? What was it about each of these events that caused the change? How did you think and feel before each such event?**

- **Can you see that you have been following mental programming that you received when you were very young and that your entire life has been consistent with that programming?**

- **Do you believe that you can change your programming? Why or why not?**

- **What specifically triggered your purchase of** *The Enlightened Savage?*

- **How did this trigger impact you?**

- **Why did you respond differently to this trigger than you may have responded to other triggers in the past?**

Chapter 4

From Cause to Effect

We have just seen how your core beliefs are formed and how they affect your entire life and everything in it. This chapter explores the process by which our realities come into themselves in detail.

Root Causes

Reality does not conform to the ideal, but confirms it.
Gustave Flaubert

Your core beliefs are where the process of making your personal reality real begins. Let's move on by examining the other layers of programming that connect our brains to the outer world.

This programming sits between your brain and the personal reality that you experience. We've already discussed how all software depends entirely on the underlying hardware and used the example of how an audio program installed on a computer without speakers won't produce sound. This is true because software can only function in the context of the hardware's capabilities and limitations. Our brain's chief limitation is that it's designed to keep us from being killed and eaten by predators. How does this work and how can we use this knowledge to become enlightened savages? There are some great questions.

Imagine a tree whose fruits are too small, too dry, and taste awful. Is the fruit the cause or the effect? If you guessed the effect, you're right. What's the cause? It could be almost anything from over- or under-watering to bad light, poor soil, improper fertilizer, or any number of things.

All of these problems affect the roots that anchor and nourish the tree, meaning that the roots are where the problem lies. Do whatever you want to the fruit and you won't make it any juicier, larger, or more flavorful. To fix the fruit, you must fix the roots. Good roots yield good fruits. Affect the roots even a little and you affect the entire tree. The physical world that you see, hear, touch, smell, and taste is nothing but the fruit of your inner programming.

What if You're OK?

Are you a Christian, Muslim, Buddhist, Jew, agnostic, or atheist? Are you politically conservative or liberal? Say, do you like seafood? I don't, but I probably enjoy at least one food that you don't like. Do you enjoy or dislike camping? How about flying? These are just a few examples of humanity's marvelously rich diversity. But here's the question: Who's right? Answer: Everyone! Why? Because each of us forms our own personal reality.

I'm tough, I'm ambitious, and I know exactly what I want. If that makes me a bitch, OK.
Madonna

If you are an atheist liberal who loves seafood, that is your personal reality, your truth. If your friend is a religious conservative who dislikes seafood, then that is his personal truth. You're both right.

Imagine ten identical computers. Nine of these have Word Processor A installed, and one has Word Processor B. If A is a powerful, easy to use program, users will have a great time. If B is a terrible program that crashes all the time, users will have all kinds of problems.

The computers themselves are working perfectly. A sound card is a sound card, a monitor is a monitor, a hard drive is a hard drive, and so forth. The program is the only difference. Program A is meeting or exceeding expectations. We call that success. Program B isn't. We call that failure.

Let's think about this for a moment: If we base our definitions in terms of expectations, desires, and goals versus results, then success and failure exist because you're either going to meet your expectations, live your desires, accomplish your goals, and achieve your desired results, or you're not. But from a program execution standpoint, where your brain is executing its survival beliefs, its core beliefs, to the letter, there is absolutely no such thing as failure.

This has tremendous implications. Your brain is, and has been, following your core beliefs from conception until this moment, and will continue following them for the rest of your life or until you change them. It is executing your mental programming to the letter. This can only mean that you are successfully following your beliefs and creating your personal reality based on them! If your program is set for wealth and happiness, then your life is rich and happy. If your program is set for frustration and defeat, then you're living one tough life.

Here again, we see how materialist evolutionary science dovetails perfectly with quantum physics. Both lead to the same predictions about how our beliefs color and shape our lives.

There May Be Nothing Wrong

It's no good talking to a man with an apology for a brain.
Lester Cole

Let's return to the tenth computer with the lousy program. What do you think might happen if you remove the bad word processing program and install the good one? The computer would execute the new program just as well as it executed the old program.

Since Word Processor A is such a great word processor, the tenth computer will be doing the same thing the other nine have been doing all along. Why? Because the computer hardware is working perfectly. The bad program is the problem because the computer has no choice but to follow its programming exactly as written.

The fact that the computer was able to run the bad program means that it can run the good program. In other words, the fact that your life does not have all the wealth and happiness you want very probably means that your brain is working properly. All we have to do is change your programming! Yes, you read me correctly: Your brain has successfully followed your core beliefs, your inner programming. This is a good thing!

You Can Change

Change your existing core beliefs and your results, your world, your entire personal reality will also change. Don't believe me? Let's look at how this works in more detail. We'll begin actually changing your programming in later chapters. I will say this much now: Making this change is one of the easiest things you're ever going to do. Paradoxically, it's also one of the most difficult.

Personal change, growth, development, identity formation—these tasks that once were thought to belong to childhood and adolescence alone now are recognized as part of adult life as well.
Lillian Breslow Rubin

Layers of Programming

If the brain is the hardware, then your core beliefs are among its earliest and deepest software entries along with your earliest emotional programming that created them. These beliefs form our brain's equivalent of a computer's BIOS, which connects higher-level software to the hardware.

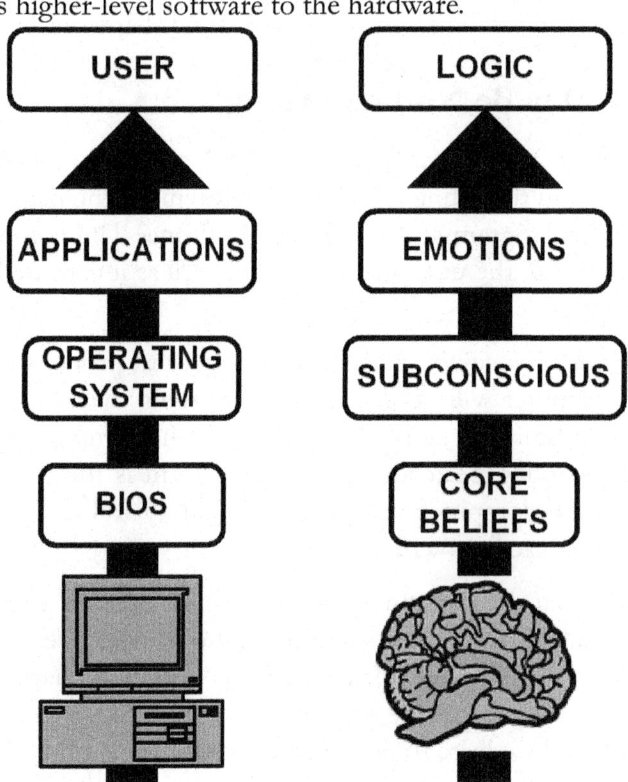

A computer's next layer of software is the operating system that connects programs such as word processors to the BIOS. In our brains, the operating system consists of our subconscious thoughts and feelings. Remember that our brain takes in 400 billion bits of information each second, while only being consciously aware of 2000.

A computer's programs include word processors, spreadsheets, games, etc. The human equivalent is our conscious emotions. Emotions have nothing whatsoever to do with logic. If you've ever known something rationally while still feeling something contradictory, you know exactly what I mean.

Reprogramming the unconscious beliefs that block fuller awareness of creative/intuitive capabilities depends upon a key characteristic of the mind, namely that it responds to what is vividly imagined as though it were real experience.
Willis Harman

The user is the final layer. Inside our brains, our logic and rational thinking form the top layer.

Each layer depends on the underlying layers. The user can't play a game unless it's installed on the computer, the game won't run if the operating system won't support it, the BIOS won't connect it to non-existent hardware, and so forth. The deeper the layer, the less direct control we exert, but the more control each succeeding layer exerts over us. Talk about paradoxical.

The key to getting the most from our computers is knowing how the hardware and software work and upgrading as needed. We can't upgrade our brains, nor do we need to. We can, however, upgrade the BIOS, which will have an immense effect on every one of our succeeding software layers.

I hope you understand that I am making the comparisons between evolution and computers because computers are a near-perfect analogy for our brains, beliefs, emotions, and thoughts. I cannot stress enough that I am neither saying nor implying that humans are robots; far from it. We'll get into that more later.

The particular element in each manifestation comes from the emotions: and just as we have our own particular emotions, so we have our own beauty.
Charles Beaudelaire

The Process of Realization

How do our core beliefs create our personal reality? I already explained how sunglasses alter light before we perceive it. Put on yellow sunglasses and you'll see the world as yellow, and so on. Never forget that sunglasses only alter our perception of the world, not the world itself. Replace the yellow sunglasses with pink ones and the world will look pink—a perfect analogy to describe what happens when we change our programming. Let's look at the ETEAR process of creating personal reality.

Emotional Programming

All learning has an emotional base.
Plato

Children lack rational and analytical abilities and must therefore rely on emotions. These emotions may begin at the moment of conception. Medical studies show that people experiencing stress, depression, anger, etc. generate toxins within their bodies. That's right: Negative emotions are literally poisonous! If the mother is generating these toxins, it's quite possible that the unborn fetus is receiving these same toxins.

If this is true, then this little human being inside the womb may already be experiencing negative emotions that will stick with her or him until she changes them or dies, whichever comes first. Look around you. Negativity is everywhere and comes in thousands of forms. Why? Why is there so much angst and suffering in the world?

The answer goes right back to the fundamental concept of humans as prey animals who are always in danger and who must always be vigilant for predators waiting to attack from all sides. This constant anticipation of terrible things is extremely traumatic. Just ask any soldier who has been on a combat patrol. The waiting and the uncertainty are almost always worse than the actual fight. I can only imagine that the same idea holds true for prey animals in the wild.

This idea explains why humans are so easily depressed, so easily stressed, so easily worried, and so quick to focus on the negative. Pregnant mothers being no exception (and in fact even more prone to emotional swings), the fetus may well be absorbing all that negativity, powerless even to realize that it's not coming from himself, because he cannot distinguish himself as separate from his environment.

The child is born and his brain immediately begins absorbing everything around him, storing this data as survival instructions in perfect accordance with its primary mission to prevent him from being killed and eaten by predators. Never mind that no predator will ever threaten this child because, again, civilization has advanced far too rapidly for our brains' programming to keep up. As far the child's brain is concerned, he was born in the middle of the Serengeti in Africa surrounded by lions, pythons, and eagles. So primed, this child begins feeling his way into the world. Have you have ever heard the expression "Do what feels right?" That's more true than you probably ever imagined.

Core Beliefs

Belief, like any other moving body, follows the path of least resistance.
Samuel Butler

As the child grows, his logical, rational, and analytical powers start developing on top of his emotions. The emotional programming he has been receiving leads to conclusions, which I call the core beliefs. These beliefs may be conscious or unconscious. Either way, each of us interprets every one of our core beliefs as "the truth." For example: "I am being beaten, therefore I am inferior." "I am being ignored, therefore I do not deserve to be heard." "My parents are extremely stressed and extremely worried, therefore the world is someplace to fear."

Do you see how easy it is for emotional programming to form conclusions, especially when you cannot separate yourself from the programming because it's not something that you are witnessing but something that is happening to you?

The moment a conclusion is formed, it becomes Truth. If your parents shoved you into the burrow at midday, then your predators must be out looking for you at noon. That is your truth, your personal reality.

Feeding the Addiction

Initial emotional stimuli (programming) form our core beliefs. These core beliefs turn right back around and generate more emotions. In very general terms, this occurs because emotions stem from chemicals generated in the brain and distributed to the body through the bloodstream. Each cell in our bodies has receptors for these chemicals, and each cell responds whenever a chemical docks with its designated receptors. In other words, emotional states literally affect our entire being at a cellular level.

Let me phrase this a different way: Your brain produces different chemicals for different emotions, and the cells in your body have receptors for these chemicals. This means that everything you feel actually changes your body's chemistry.

Think of any addictive drug: When the user first begins taking the drug, a small dose provides a nice high. But then what happens? Soon the user must start increasing the dose in order to feel the same high. Any attempt to stop using the drug causes the entire body to experience withdrawal symptoms because it has become dependent on (or addicted to) that drug, and requires it for normal functioning. Drug withdrawal can be a long agonizing process that continues until the body resets its chemical processes.

Emotional chemicals work the same way. Experience the same emotion often enough, and your body may develop an addiction to the chemical (or chemicals) associated with that emotion. In other words, you might literally become addicted to that emotion. Like any good junkie, your brain will do everything it can to justify and continue that addiction because it is based on a logical conclusion, on "Truth." In turn, this logical conclusion stems from your emotional programming and experiences that you could not differentiate as being separate and distinct from yourself. The idea that your core beliefs are your Truth stands to reason, because each of us creates our own personal reality, and that personal reality must be based on something.

If we can be addicted to heroin, we can be addicted to any neuropeptide (emotion). What is thought of as "reality just happening to us" is really a result of consistent choices producing specific chemicals, which result in specific emotions that have become a habit.
Dr. Candace Pert
(paraphrased.)

If you see yourself as worthless, scared, unfulfilled, etc. then your emotions will play right along. As awful as those emotions might be, you're going to stick with them because they are a direct response to the survival instructions you received in order to keep from getting killed and eaten! The addiction you develop to those emotions will keep you coming right back for more and more, thus creating the perfect self-fulfilling prophesy.

These feelings are always whispering to you, saying "This is okay. It's safe to do. This is what I must do and this isn't." Right? No way you'll voluntarily leave that comfort zone, because leaving the comfort zone generates the fear of being killed and eaten. Make no mistake: Each of us is being led around by our emotions, which use far more of our brain's capacity than our logic.

Here's an example: I once spoke with a woman who said that she wanted to marry a wonderful man and raise children in a nice house. Her profession? Prostitute. Why? I believe it's because she believes that she isn't worthy to lead that kind of life. I explained the process of realization to her just like I'm doing with you. "You're right," she responded, "but I'm not ready to change." How's that for the power of negative core beliefs and emotional addiction?

Once you fully understand that you have control over your thoughts, you can choose thoughts that will create emotions that will give you a "high" everyday!
Teri Hoskins

Let's examine this story a little more closely. She wants to marry a wonderful man, but her actions are dooming her to failure. This much is blindingly obvious; one must wonder why she chose this combination. The answer lies in the fact that the conscious mind only comprises a tiny fraction of our brain's total functions. I believe that the core beliefs in her unconscious mind deem her unworthy and deserving of defeat. Her brain doesn't want her to get killed and eaten any more than yours wants that fate to happen to you. Thus the die is cast. On a conscious level, she is failing miserably at achieving her goal. On a subconscious level, however, she is succeeding perfectly—that is, if you accept my theory that most of our struggles and tribulations are caused not by illness but because our brains are perfectly healthy.

I am no doctor and so cannot render any sort of medical diagnosis. That said, I believe that most of our problems stem from successfully executing flawed programming. In other words, I believe there is far more right with most of us than there is wrong.

Bridging the Gap

The most decisive actions of our life—I mean those that are most likely to decide the whole course of our future—are, more often than not, unconsidered.
André Gide

Remember that our emotional programming leads to core beliefs that feed emotions and emotional addictions. So far, however, nothing has happened. The next step in the process of realization must therefore be action.

Actions are what bridge the gap between ourselves and the outer world. For every action, there is a result. If you believe that you are not worthy of being married to a great guy and raising kids in a nice house, you will take steps to prevent that reality from occurring. Being an escort is a pretty good way to fulfill negative core beliefs by keeping those dreams from coming true. The conscious goal exists to serve the subconscious programming that long ago slid into the driver's seat.

Creating Our Personal Reality

To recap, the ETEAR process of realization is how each of us creates our own personal reality. The process works as follows:

Creation is a drug I can't do without.
Cecil B. De Mille

1. Emotional core beliefs (E) lead to thoughts (core beliefs)
2. Core beliefs (T) drive emotions and emotional addictions (E).
3. Emotions direct actions (A).
4. Our actions create our personal reality (R).

As you can see, each of us does indeed create our own personal reality; however, this is not the same thing as "attracting" or "manifesting" things into your life. Every person's personal reality exists as both a subset and a superset of shared reality. Shared reality must always conform to the laws of physics. As we will see in Chapter 23, the universe does have a little wiggle room—but not much. This shared reality is why the "Law of Attraction" does not and cannot work.

How I Learned

We already know that humans are prey animals and thus prone to vigilance, stress, and worry that the other shoe is about to drop. Given how easy it is to saddle ourselves with self-destructive programming and beliefs, is it any wonder that every new generation inherits that negativity from their parents?

There is always one moment in childhood when the door opens and lets the future in.
Deepak Chopra

Speaking of learning, people learn in three primary ways:

- Hearing, such as from parents, clergy, teachers, or news telling them things.
- Seeing, such as witnessing how parents interact.
- Specific events that affect them, including actions.

Of the three, specific events are by far the most insidious since they can have a profound effect on one person's life while seeming trivial to other people.

Here is an example from my own life: When I was about six years old, I really wanted a toy microscope. My parents were poor in those days, but my mother said she'd be glad to buy it for me. She later told me that they didn't have the money and couldn't buy it because we needed food. Boy was I angry! My mom had broken a promise. I confronted her and she kept repeating that she just couldn't afford it. Finally I gave up. But then what happened? She bought the microscope.

You'd think I would have been ecstatic, but that's not how I felt. This event filled me with guilt because I believed that my parents had gone hungry for my stupid toy. I vow never to feel that kind of guilt again and vow never to deprive myself of anything. Money? It's a weapon in a world where anyone lacking weapons is at a serious disadvantage. I started buying every toy in sight the moment I began earning money. New cars, houses, computers, DVDs, and more, as much as I could get my hands on—and more.

Creation is going on all the time.
George Bernard Shaw

All this happened because I was trying to satisfy this six-year-old child who was profoundly affected by what, in retrospect, was a trivial thing. Mention your experience and its effect to the other person who was there and chances are they'll be blown away.

I was 35 when I finally told my mom about the microscope episode. Stunned, she told me that she really wanted to buy it but wouldn't be able to buy groceries if she did. Then she got a little bonus in her paycheck and immediately bought me the microscope. She acted out of love and caring for my real needs, and I interpreted it as her using money against me. The kicker is how that one incident affected me for over thirty years. Five years after that talk with my mother, my entire house of cards came crashing down around my ears when the economy tanked in 2008 because the lesson just hadn't gotten through to me in time. From husband, homeowner, and breadwinner, I became divorced, bankrupt, and unable to pay for my house or lifestyle.

Don't laugh, folks, because you probably have at least one of these experiences in your own past. Again, this incident occurred when I was six. Why did it affect me so powerfully? Why did I not interpret it as an act of love? Because I was jam-packed with negative core beliefs and was building my personal reality in accordance with those beliefs.

No "Law of Attraction"

Thousands of self-professed self-help gurus have put out tens of thousands of seminars, workshops, books, CDs, DVDs, and more promoting the notion of the "Law of Attraction," a concept that originated in "New Thought" and theosophical writings of the 19th and early 20th centuries. The basic idea is that positive and negative thinking cause corresponding results, and that one can "attract" whatever one wants by simply adopting the proper thoughts and beliefs. This much is true, albeit not for the reasons its adherents claim.

We've always been here and we'll always be here. We are a specific arrangement of particles and this instant is infinite. Did we luck out, or didn't we? The odds against this sentence having ever being typed, much less the odds against you reading it were inconceivable. Smile, because the fact that you're able to is almost impossible to comprehend.
Jeffrey Rowland

If the "Law of Attraction" could be used for any practical purpose at all, we should expect that mainstream modern science would have amassed empirical proof of its existence. No such thing has occurred. On the contrary, mainstream science sneers at much of the self-help industry, and with very good reason: It doesn't work. This is not to say that the self-help industry is bad; on the contrary, it has a lot of good ideas that have a lot of truth to them; however, these ideas don't work for the reasons their proponents think they do. Understanding this makes all the difference between an otherwise decent concept that does not pass scientific muster and one that does. As you continue reading this book, you will come across ideas that sound very similar to the "Law of Attraction," but they are in fact subtly yet powerfully different. More in Chapter 23.

Summary

The ETEAR process works as follows: Emotional programming (E) creates logical core beliefs (T) that spawn the feelings (E) that drive the actions (A) that generate our results (R). This process may begin before birth, because bodies under stress release chemicals which pass to the fetus. If so, the infant is born primed and ready to accept its programming. If the mother was stressed, worried, nervous, or angry, the newborn infant may be more apt to filter everything it sees from then on in those terms. If no emotional chemicals pass from mother to infant, and no programming occurs during pregnancy, then the programming begins at birth. It's important to remember that this is a survival mechanism.

Your brain uses its core beliefs as survival instructions, quite literally, "I must act this way in order to avoid getting killed and eaten by my predators." The fact that humans have survived this long is a testament to the effectiveness of our survival instincts. Think about this: Each of us is the result of an unbroken chain of ancestors going back millions of years. One break in my chain, and I wouldn't be here. Same for you.

I firmly believe that a good measure of human problems and suffering come from the misguided idea that people are a third kingdom of life separate from plants and animals—a pretty arrogant stance coming from a breed of diseased monkeys! I for one find the notion of humans as diseased monkeys pretty humbling. This disease, this mutated gene that shrank our jaws and allowed our brains to grow, led to civilization, a thin coat of paint covering a large evolutionary brick.

So now you know. You know exactly how you've gotten yourself into the pickle the prompted you to buy this book. You may feel like a total failure. Facing bankruptcy and foreclosure, I know I sure felt that way. But the thing is, my situation is not a failure and neither is yours. On the contrary, my entire life is a string of 100% success right up and including my personal and financial meltdown. So is yours. Once you understand that, you can begin redefining what success really means to you. Once you do that, your successes will become the kind of successes you want, not the kind that seem to just keep happening to you. This process bears superficial similarity to—but is very different from—the "Law of Attraction." (See Chapter 23 for more on why this is so.)

Questions

What do you think your core beliefs and emotional programs are? Understanding this programming is a critical step to upgrading it. Write down your answers to the following questions:

- **Do you accept the fact that you create your own personal reality and everything in it? If not, why not?**
- **Do you understand that you must accept responsibility for the life you have in order to own your power to make the changes you want?**
- **Are you willing to undergo some very fundamental changes?**
- **Do you understand that these changes may feel very uncomfortable at first?**
- **Are you committed to living and growing through the discomfort in order to experience the life you want and deserve to live? If not, why not?**
- **Is there one seemingly unimportant episode in your youth that might be continuing to influence you today? Describe it in as much detail as possible.**
- **What if your core beliefs were different? How do you think that might have altered your life to date?**

Chapter 5

The Truth about Failure

In this chapter, we'll look at how your core beliefs are acting against your wants and needs and begin laying the foundation for lasting transformational change in your life.

Redefining Work

We work to eat to get the strength to work to eat to get the strength to work to eat to get the strength to work to eat to get the strength to work.
John Dos Passos

Question: Are you a hard worker? Do you equate "success" with "hard work?" If so, then you have fallen into one of the most insidious traps ever created, one that is literally draining away your life. To illustrate how this trap works, I like to use the example of two identical piles of dirt that must each be moved an identical distance.

Person A tackles the job using a tiny kiddie shovel and bucket and completes the job after days of backbreaking labor. Person B hops on a tractor and moves the entire pile in one scoop that takes less than a minute. Who did more work? Person A was out there for days, while Person B never broke a sweat, so Person A must have done far more work, right? Wrong. They each did the exact same amount of work!

Each is doing the same amount of work.

There lies the trap because most people confuse "work" with "struggle." In other words, they confuse the end result with the effort required to achieve that result. Physics, however, defines *work* as a measure of *performance*, or output. I'll say this again: Work is not a measure of effort, but of results. Person A struggled to achieve the exact same result that Person B accomplished with ease. Therefore, they both did the exact same amount of work.

Hard work. Well, that's all right for people who don't know how to do anything else.
Dan Totheroth

This simple fact goes to the core of why most of us never achieve the success we are destined for. Society conditions us to define "work" as "struggle," with the result that we're expected to spend 40+ years in a job. Why? To scrape together enough money to live out our few remaining years in some semblance of comfort before shuffling off this mortal coil. If that isn't slavery, then I have no idea what is.

Self-employed? Guess what: If your business requires your presence the majority of the time in order to keep revenue coming in, then you are even more of a slave than the millions of worker drones who can quit and find a new job any time. Walking away from a job is one thing, but walking away from a business isn't quite so simple. If you're in business, you have the choice to design (or redesign) your business to either give you the freedom and resources you need to achieve your life's goals or make you a slave. It's up to you. Sadly, most people resign themselves to slavery and blame anyone and anything but themselves for their predicaments because they insist on defining work as effort instead of results.

It is a wise person that adapts themselves to all contingencies; it's the fool who always struggles like a swimmer against the current.
Anonymous

Think about the people you know. Do some of them have the magic touch where everything they involve themselves in blossoms and rewards them? Are others the kiss of death to jobs, businesses, investments, etc.?

Why is that? Are some people lucky and others unlucky? Does God, the universe, or karma favor some individuals more than others? No. Each one of us has the exact same potential for success and the exact same chance to achieve everything we want in life, and then some. I know children from poor refugee families who are living very comfortable lives. I also know people from wealthy families who are bumbling aimlessly from day to day. So what's the difference? What separates successful people from failures?

I hope you guessed "negative core beliefs," because that's the answer. Those beliefs are the roots, and our lives are the fruits. If you want to get better effects, you must tend to the cause by replacing those negative core beliefs with positive ones. If you don't believe that personal reality comes from within, then how, precisely, do you explain the fact that 20 witnesses to the same event will give you 20 different descriptions? If you believe there are voices whispering inside your head, then those voices are just as real to you as not hearing them is real to others.

Work in the "Real World"

The pleasure of life is according to the man that lives it, and not according to the work or the place.
Ralph Waldo Emerson

Let's apply this to the kind of life far too many of us lead. Society wants us to believe that the road to success requires struggling at a job for 40+ years. Buy into that belief, and you'll act on it by going out and getting a job, or by creating your own job by starting a business. Forty-some years later, tired and spent, you'll have achieved the expected result: a short time of relative leisure before death. Die too soon, and you'll leave behind part of the resources you spent your entire life amassing—an utter waste. Live too long, and you'll expend your carefully built resources, only to die with nothing whatsoever to show for your lifetime of slavery. I ask you, is that prospect depressing or depressing?

The Cycle Continues

Remember the process of realization from the previous chapter? The vast majority of us receive at least some negative core programming. We conclude that survival depends on living up to people's expectations because we're not worthy of being free individuals. As social animals, we do our utmost to fit into the group. We become addicted to that emotional pattern and enter the rat race with absolutely predictable results. Oh sure, we sometimes dream of something different and better, and maybe even buy a self-help program and/or go into therapy.

Work to survive, survive by consuming, survive to consume: the hellish cycle is complete.
Raoul Vaneigem

All too often, however, we twist these programs and therapy to fit our core beliefs and soldier on, faithful to the bitter end. Take a good look around you. Whatever you have, whatever cycles you're locked into, will remain the same for the rest of your life. You may as well get used to this idea.

Remember that we learn by seeing, hearing, and experiencing. What did your parents, teachers, friends, clergy, associates, bosses, the media, etc. tell you about money, happiness, wealth, and success? How did your parents view work, struggle, success, money, and fulfillment? Were you told that rich people are greedy crooks or that you can't have everything you want?

Were you abused? Pushed to do better no matter how well you did? Most importantly, how do (or did) your parents and those closest to you live their lives and what are the results? What specific life experiences, no matter how seemingly minor, affected your beliefs? And—most importantly—when were these beliefs formed? Remember that core beliefs form between the time of conception and roughly age two, with everything that comes after adding itself to the heap.

If you believe that success requires struggle, you will struggle hard. If you believe that rich people are greedy criminals, you will never be rich. If you believe that money is the root of all evil, then chances are you're either poor or perilously close to poor. If your parents treated money with anger, chances are you associate money with anger. Same with happiness, fulfillment, purpose, etc. By now you know that you never believe things when you see them; you see things because you believe them.

This cycle is compounded and reinforced by everyone around us, because humans are social animals that evolved living in groups. Everything you see, hear, and experience tells you what the group (society) expects of its members. Your core beliefs help you decide your place in the group. Fitting into the group is essential, because it offers safety and protection from predators and rivals. To be cast out of a group is one of the most traumatic experiences one can have. We will therefore do all in our power to follow our survival instructions and fit into the group.

A Different View

Every adult, whether he is a follower or a leader, a member of a mass or of an elite, was once a child. He was once small. A sense of smallness forms a substratum in his mind, ineradicably. His triumphs will be measured against this smallness, his defeats will substantiate it.
Erik H. Erickson

What if you saw some of the things you saw as a child today for the very first time? Think your reaction might be vastly different? I sure hope so! As adults, we understand that we have the power to interpret what we see, hear, and experience in many different ways. Take two people on the same tropical vacation. One will love the exercise and the wildlife, the other will complain endlessly about schlepping through some God-forsaken jungle being eaten alive by mosquitoes and leeches. But children don't have the capacity to envision multiple interpretations. They make one interpretation and that's it.

We've already explored how beliefs work using the sunglasses example. Let's recap very quickly: Put on yellow sunglasses and the world appears yellow. You know that the world is not yellow; it only appears that way because the glasses are filtering the light before it reaches your eyes. Your beliefs are nothing more than sunglasses. Wear sunglasses for a few minutes and you'll forget you have them on, but they're still there coloring your world. So are your core beliefs.

Everything in life that we really accept undergoes a change. So suffering must become Love. That is the mystery.
Katherine Mansfield

Breaking the Cycle

Changing your life is as simple as changing your programming. Getting great results without great effort is as easy as upgrading your definition of work. How can you go about doing this? Answering that question is the whole purpose behind *The*

Enlightened Savage. Even better, the fact that you're reading this book means that you have already taken those all-important first steps.

Identifying Your Programming

Identifying your core beliefs is the first step in changing them. What are your beliefs around wealth, happiness, self-worth, and success? Are these beliefs contributing to your life or detracting from it? If they are detracting from it, the good news is that changing them is just as easy as changing sunglasses.

The cause is hidden; the effect is visible to all.
Publius Ovidius Naso

Hidden in Plain View

How the heck can you possibly identify your core beliefs if they're locked in your subconscious mind? This may seem difficult but is actually extremely easy. How do you live your life? Your causes create your effects, and your effects give you all the clues you need to expose the causes. Since you only have one set of core beliefs, you only have one set of causes governing your entire life.

The mere existence of conscience, that faculty of which people prate so much nowadays, and are so ignorantly proud, is a sign of our imperfect development. It must be merged in instinct before we become fine.
Oscar Wilde

In other words, what exists in one part of your life exists in all parts of your life. Do you take charge or do you hang back out of fear or feelings of inadequacy? As you read this book, are you finding yourself saying that I'm full of bull, spouting psychobabble, trying to sell you something, giving false promises, or even lying to you?

Think about your responses to what you're reading as you read it. Pay very close attention to how your mind is responding to this book, then apply that same response to your whole life. Let me perfectly blunt: How you respond to what I'm saying or any part of the Enlightened Savage program is a microcosm of how you respond to everything in your life.

Those responses, your entire modus operandi, is what is shaping your life and causing the effects that are your personal reality. If you are not experiencing all the wealth, happiness, joy, love, etc. in your life that you want, it is because your responses are directly tied to (and stem from) your core beliefs. Simple observation reveals these beliefs which have been lying hidden in plain view your entire life.

Wait just a minute. You're an intelligent being, certainly smart enough to grasp what I'm saying on some fundamental level and make sense of it. Why then aren't you living the kind of life you want? Remember that your brain processes some 200 million (200,000,000) bits of information for every 1 you're consciously aware of.

You are a prey animal whose brain is designed for self-preservation and self-preservation only. Your conscious mind is just tagging along for the ride of its life—literally.

But You Know That

You know nothing till you know all; which is the reason we never know anything.
Herman Melville

Are you finding yourself saying "I already know that" as you read this book? In the introduction, I explained that these are the most destructive four words in any language. The moment you say "I already know that," you shut your mind to whatever is being offered to you. Same goes for any of other restrictive and destructive response you can come up with. *The Enlightened Savage* materials that you just spent your hard-earned money to purchase can be a catalyst for sweeping change in your life, Anthony spouting trite nonsense, or anywhere in between. It's completely up to you.

Fortunately, getting around the "I already know it" block is easy. The moment you find yourself closing yourself off, simply stop and say "Thanks for the step up," then proceed ahead with double the determination. Remember that as far as your brain is concerned, any deviation from your routine means getting killed and eaten. It therefore tries its best to stop you whenever you approach the comfort zone boundaries defined and enforced by your beliefs.

It's nothing personal. Millions of years of evolution have taught your brain that it must be right in order to carry out its mission of self-preservation. It will duly throw huge obstacles in your path whenever you try to break the mold you made for yourself.

Second of all, when you say "I already know that," how can you be really sure you in fact know? This is also very easy: Look at the effects. Are your effects exactly the way you want them? If so, great! But if not, then you are surrendering to the obstacles your brain is placing in your path. The moral of the story is clear: If you already live it, then you already know it. If not, then the only person you're fooling is yourself. Look at your realty. It cannot lie, because it is nothing more or less than the exact result of the causes that have taken up residence inside your mind since the day you were born.

Rising to the Challenge

Your response to the obstacles your brain erects in your path are what will determine your success or failure in getting the wealth, happiness, and fulfillment that you want and deserve. Successful people, rich people, happy people all share one trait: They have the same fears, insecurities, doubts, and destructive beliefs as you, me, or anyone. But unlike their unsuccessful brethren, successful people press on no matter what it takes.

It is very rare that you meet with obstacles in this world which the humblest man has not faculties to surmount.
Henry David Thoreau

They confront their self-made obstacles and tell themselves "Hey, thanks for the step up!" as they turn those obstacles into stepping stones towards their next success. Be open—truly open—to success, and success will find you.

I had a client who wanted to start a business. She told me that her parents and friends were warning her against it because "it's too hard." I asked her some very simple questions: Who are these people? Are they rich? Are they living the lives they want or are they struggling? Lo and behold, the naysayers are in the same struggling boat my friend is in. Of course they're warning her off!

The lesson is clear: If you need a teacher, find someone who is doing what it is you want to do. How sensible is that? If you want to be rich, learn from a rich person. If you want to learn sales, learn from the best salesman you can find. And always learn from people whose beliefs challenge your own. Otherwise, all you're doing is reinforcing your existing beliefs. Correct me if I'm wrong, but aren't your existing beliefs precisely what landed you in your current predicament? This is why you must learn from someone who has been where you're going.

You will need to search out new ways of expressing strength, showing mastery, and exhibiting courage—ways that do not depend upon confronting the world before you as an adversary.
Kent Nerburn

Well? Is your mind saying "I already know that" and closing you off? Are you feeling like I'm selling you something, or misleading you, or making false promises, or spewing bull manure? Write each reaction you have down the moment you have it, then go back and look at your list. Right there, in your hands, you'll have the clues to what's been holding you back from the success you crave—and the success you deserve, which is often far more than your wildest craving.

Master Each Step

Master of the universe but not of myself, I am the only rebel against my absolute power.
Pierre Corneille

OK, so you realize that your results could use some improvement and are ready to take the next step towards achieving success. You've cast your old beliefs aside and are just waiting for the good stuff to rain down on you. You wait and wait, but nothing happens. Why not?

Pretend you have a toddler in your house. He's accustomed to drinking water from those spill-proof plastic kiddie cups. One evening at dinner, he decides he really wants to drink water like a grown-up. Being the affable parent that you are, you unscrew the cap and give Junior his first "big boy" cup of water. Toddlers being toddlers, he drops the cup, sending water everywhere. What do you do? Chances are, you'd get up to clean the spill, then go into the kitchen to fill the cup again. After all, Junior is just learning; besides, even the best of us knocks the occasional glass over.

You fetch towels and come back to discover that Junior has lunged up from his highchair and grabbed your water cup, which is both much larger and made of glass. What do you do? Snatch it away as quickly as possible, of course! Why? Because Junior just demonstrated that he can't handle a small plastic cup. There's no way he'll manage a bigger cup made of glass. Besides, if he breaks your glass, he could get some pretty nasty cuts. The lesson here is obvious: As a kind, loving adult, you don't want to set Junior up for failure by giving him more than he can handle. And you certainly don't want his over-exuberance to hurt him.

The Universe, God, karma, or whatever you believe in operates in exactly the same way: You will only get as much as you can manage at any one time; not a smidgen more. Wondering why your business is failing or why you never have enough money, happiness, or other blessings in your life? Because you can't handle what little you do have, and therefore can't possibly manage more. Furthermore, as bad as your situation might be, imagine how much worse you'd feel if you suddenly got everything you ever wanted and lost it all. My guess is that would cut you pretty deep—just like the shards from the broken glass could cut Junior pretty deep.

The Loving Universe

Try this concept on for size: The Universe, or whatever you believe in, is the kindest, most benevolent, and loving parent there can possibly be. Don't believe me? Take a moment to consider the staggering odds against your being born at all. One tiny slip in your lineage over the billions of years life has existed on this planet, and you would not be here to read this book. Now consider that in the 13.7 billion (13,700,000,000) or so years this Universe has existed, there has never been a creature just like you, nor will there ever be again (at least not in your current form).

Knowing that you're lovable helps you to love more. Knowing that you're important helps you to make a difference to to others. Knowing that you are capable empowers you to create more. Knowing that you're valuable and that you have a special place in the universe is a serene spiritual joy in itself.
Louise Hart

Each and every one of us is therefore a priceless, unique jewel in this grand scheme. If that realization doesn't inspire you, I don't know what will. This makes any failure to live up to your full potential literally a failure of cosmic proportions. But that is exactly what will happen, because the Universe loves you too much to set you up for failure by giving you more than you can safely manage. That safety valve is built into the deepest recesses of your brain and is constantly looking for any excuse not to get deep cuts from literal or figurative glass shards.

You must do two things if you want to succeed in business, love, finance, or anything:

1. Shed your destructive beliefs, and replace them with affirming positive beliefs that will help nurture you and pave the way for your success.

2. Learn to manage what you have. Live within your means. Know where your money is coming from and where it is going. Account for every cent. Begin saving some of your money. I realize some of you might be in bad financial shape. We'll talk about how to start digging yourself out later on in this book. For now, just rest assured that you can do something about your situation, no matter how dire it may seem right now.

Committing to Change

Succeeding requires the ultimate in patience, common sense, commitment, humor, tact, love, wisdom, awareness, and knowledge.
Virginia Satir

This book is jam-packed full of tools that can totally change your life, starting the moment you began reading. Will your life suddenly change direction and shower you with the Universe's bounty? No. That's right: I've invested thousands of hours researching and writing, and it won't do anything at all for you. The excitement and empowerment you feel while reading this book will fade away the moment you run out of pages.

The Enlightened Savage will end up in your growing collection of self-help and success books, tapes, disks, weekend retreats, etc. Why? Because Anthony's selling snake oil. Anthony's lying. Anthony's using cheap psychological tricks. Everyone knows that success requires tons of work; it never just happens. The list of reasons goes on and on. And you know what? You're absolutely right! In fact, you are so right that you should fire up your computer and go to my Web site right now to demand a refund, because your life will never change, and the priceless gem that is you will only go to waste.

What is truth?
Pontius Pilate

How do I know this? Because I know exactly what you're thinking. I know your mind is saying the same self-defeating things it's been saying your whole life. So what's the point? Why bother? No reason. None, that is, except that maybe this time you'll take all of the objections and excuses your mind is throwing out and say "Thanks for the step up!" Maybe—just maybe—this time your obstacles will become stepping stones towards a whole new future instead of deterrents.

Truth is a delusion.
Friedrich Dürrenmatt

Way back at the beginning of this book, I told you that *The Enlightened Savage* isn't *the* truth. It's *a* truth. It's a model of my personal reality. All I can do is offer you this model for you to accept or reject as you alone see fit. I can, however, assure you of one thing: I was once trapped in my own self-made prison, just like you. I struggled every day, and still came very close to being broke every month. I went bankrupt and lost my home and marriage. That's what it took for me to finally understand the lessons I'm trying to impart in this book. The day I started understanding and applying these lessons, my began changing because I faced my obstacles, and used them to my advantage instead of surrendering to them. I did so using the exact same model that I am presenting to you.

Is my transformation complete? I doubt it ever can be. But I can either get busy living or get busy dying, to quote Stephen King in *The Shawshank Redemption*. Being an enlightened savage isn't about the goal; it's the journey that counts.

Breaking Through

What is the most rigorous law of our being? Growth. No smallest atom of our moral, mental, or physical structure can stand still a year. It grows—it must grow; nothing can prevent it.
Samuel Clemens

A few paragraphs back, I remarked that even the happiest, richest, and most fulfilled person on Earth has doubts, fears, and insecurities. I'll let you in on a secret: Even now, my brain is screaming at me to close my word processor. I'm an awful writer and the material is worthless, *The Enlightened Savage* is nothing but a colossal waste of my time and money, and so on. And you know what my reply is? Thanks for the step up! Thank you, brain, for making me so uncomfortable that I know I'm doing the right thing.

Here's another dirty little secret: The fastest, easiest path to personal growth and learning is the path of most resistance! If there's a wall in front of you, is it faster to way miles out of your way to go around, or to punch straight through? Hey, if you want the easy way out, then go find yourself a therapist who will pump you full of medication and spend years asking you how you feel about every piddling memory you have.

The shortest path is straight through the obstacle.

A quick reminder: There are psychological and psychiatric disorders that require intervention by trained, certified mental health practitioners. Do not construe anything in this book as offering either medical advice or encouragement not to seek any form of treatment.

It takes courage to push yourself to places that you have never been before... to test your limits... to break through barriers.
Anais Nin

So do you want the long, slow, comfortable path, or do you want the fast, effective way? If you're ready to get on with your life, then break on through the walls of your self-made prison to the other side. I promise you that the grass on the other side is a whole new shade of green. My friend Jay Conrad Levinson did it and redefined the way the world does business. Another friend and mentor, Jim Britt, went from making $1.67 per hour in a factory to being a multimillionaire with a wonderful family and home. I'm doing it, and every day brings new blessings that I couldn't have imagined even a week ago.

Drop the Baggage

The perfection of traveling is to travel without baggage.
Henry David Thoreau

In June of 2004, I attended a four-day workshop put on by Jay and Mr. Chet Holmes, called *Guerrilla Marketing meets Karate Master*. This workshop ran four long days, from Wednesday through Saturday.

That Saturday evening, I was hanging out with Jay, his wife Jeannie, and some of the Guerrilla Marketing coaches. I don't remember what I said at one point, but master coach Larry Loebig leaned forward, looked me in the eye, and said "Hey! No negative self-talk. You're part of the Guerrilla Marketing inner circle now, and we don't take that here." I began babbling something about my self-esteem and the work I was doing. Larry interrupted again. "Look," he said, "you're a brilliant guy. You don't need the baggage. Drop it."

I fooled myself into thinking I had dropped my baggage and all was well until I lost everything in the crash of 2008. Now I know better and am giving you the same advice that Larry gave me: Drop your baggage. You don't need it. All it's doing is weighing you down. We'll get to work identifying your baggage in the next chapter. Meanwhile, concentrate on this one idea: You are a priceless, unique human being. You deserve every success. If you decide to drop the baggage, you will get the success you want. Just make sure that the baggage is really gone and not just hidden. Don't make my mistake!

Questions

It is very important for you to take the time to listen to the objections your mind is throwing out, because they hold the key to your lack of success. Write them down. Once you identify why you're not achieving the success you want, the next step is to learn where your core beliefs came from and why.

- **Do you define work as effort or as output?**

- **Do you find yourself getting results with ease or do you need to struggle to accomplish your goals?**

- **Do you know any "lucky" people? If so, what makes the different from everyone else?**

- **Do you know any "unlucky" people? If so, what makes them different from everyone else?**

- **Are you doing what you want to do in life or what you feel you must do?**

- **If you could do and be anything you want, what would that be?**

- **If present trends continue, will you be able to retire comfortably?**

- **How well prepared are you for unplanned events? How long could you maintain your current lifestyle if you stopped earning an income?**

- **Looking back over your life, can you see up and down cycles of money, happiness, and fulfillment? Can you pinpoint any events that triggered these peaks and valleys?**

- **What have you been taught regarding wealth, happiness, and doing what you want in life?**

- **Do you believe that success requires "hard work"?**

- **Does your daily life reflect your true priorities?**

- **Do you know all of this already? If so, does this knowledge realize itself in your daily life? If not, why not?**

- **How well do you manage your life and finances?**

- **Do you accept and own the fact that you are a unique and priceless being who deserves every success? If not, why not?**

- **Are you feeling tempted to abandon this program? If so, why?**

- **Do you feel like this book is making a bunch of false promises? If so, have you felt this way about other programs you've tried?**

- **If your past efforts to obtain the life you want and deserve have not yielded your desired results, are you willing to try something completely different?**

- **Describe any additional doubts or concerns you may be feeling right now.**

- **Why do you think you are not living the life you want and deserve to live? Be as detailed and specific as possible.**

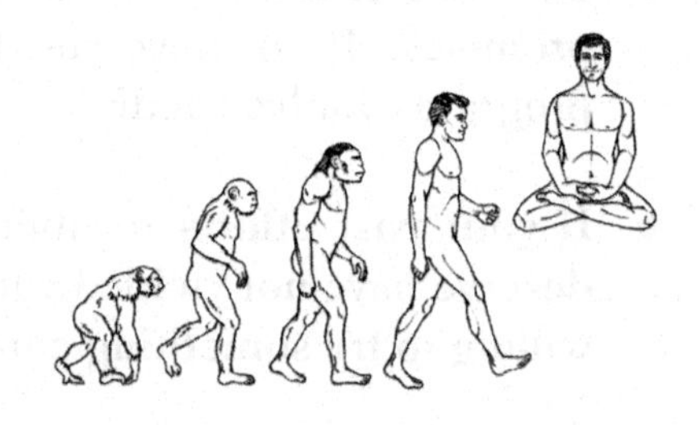

Chapter 6

Taking Stock

I consider that a man's brain originally is like a little empty attic, and you have to stock it with such furniture as you choose.
Sir Arthur Conan Doyle

Having identified your reasons for not achieving the success you want, your next step is to find out where those reasons came from and why. You'll then be well on your way to beating them, because knowing your obstacles is the first step in beating them. But before you go any further, you need to take a moment to thank yourself for your courage and perseverance in sticking with me this far.

Remember that the more discomfort you're feeling, the more you're growing. This book is starting to challenge beliefs that you've been building and reinforcing since childhood. I'm only one lone enlightened savage setting out to ambush your entire army of negative programming, and I just don't have time for subtlety. It's not that I don't respect you. On the contrary, I respect you far too much and value your limited time far too highly to waste it. You and I need every second of our limited time together to do the most good. I hope things are starting to fall into place for you and that you'll start experiencing some powerful changes very soon.

Where Did Your Beliefs Come From?

> *The real challenge (in life) is to choose, hold, and operate through intelligent, uplifting, and fully empowering beliefs.*
> Michael Sky

You know that people learn by seeing, hearing, and experiencing. What did you learn about money, love, happiness, dharma, and success?

Who did you learn this from? Who were your role models? What specific life experiences helped shape your beliefs? If you remember my microscope story, then you know how a seemingly trivial event can alter a life. Do you have a similar story? I haven't yet found anyone who doesn't, so I'm guessing that you have at least one. It's OK. Recognizing the stories is the first step to finally moving beyond them.

Let's talk about each of these methods in a little more detail. How did the important people in your life act around money? How did they define success? Did they tell you to go out and get a job? Was Dad always traveling or at the office until way past bedtime? Was money a source of joy, sadness, anger, frustration, fulfillment, frustration, or any other emotion?

What about life experiences? I'm still dealing with the repercussions of my microscope experience. I only recently got over the guilt of being convinced that my parents had decided to go without food to please me. And that stuck. So what did I do? I struggled hard for my money, putting in insane hours to keep earning more and more. My needs grew right along with my earnings. I had to have a new car every two years and went on a buying spree that only ended when I was laid off in December of 2002, with no job offers in sight. I thought I'd learned my lesson but I was only fooling myself. I failed to take enough steps to reverse the damage and walked right into the economic meltdown completely unprepared. It took me until late in 2009 to finally start taking the actions I needed to take to put my past behind me and get a fresh start. The repercussions will continue for years. Why? Because I didn't learn my lesson when I had the chance... and therefore had to face another and far more powerful lesson.

Did I bother learning about money management? What, and confront the same kinds of choices my parents did? My story illustrates the invisible, yet powerful, stranglehold that our beliefs have on our lives and our resulting personal realities. I was six years old when this event took place. Any adult can see that my parents cared for their son as best they could. Unfortunately, six-year olds aren't capable of rational adult analysis.

That experience ruled a good part of my life for decades, and I am far from alone. Every person on this planet has their own demons. Tragically, most die with them; many even die because of them.

Success Defined

Everything you need for a better future and success has already been written. And guess what? All you have to do is go to the library.
Jim Rohn

I'm using the word "success" throughout this book. What exactly is this "success" we keep talking about? Money, happiness, joy, a profitable business—There are as many ways to succeed as there are people on this Earth. I personally define success as the freedom to live, love, and laugh, and to make a powerful difference in the lives of everyone I meet. And you? How do you define success?

I also use the word "wealth." Most people define wealth as having a great deal of money. That's totally OK by me. Just don't forget that the concept of wealth goes far deeper than a balance sheet. True wealth comes from realizing your life's mission, or dharma, and setting out to accomplish that mission. Being true to your dharma is the greatest form of wealth there is. We'll talk about dharma a little later.

Make no mistake about this: I have absolutely nothing against healthy bank accounts. I simply don't limit my definition of wealth by the mere size of those accounts, and neither should you. As the Bible says, what shall it profit a man to gain the whole world and lose his own soul? Nothing. I know. I've been there, and done that. And you know what? It took my being absolutely broke with no material wealth to show for 40 years of being alive for me to finally discover how to stop passively basing my happiness on external things and start actively choosing to be happy. The truth, as we'll discover later, is that there is absolutely nothing intrinsically good or bad about anything that happens in your life.

But I'm getting ahead of myself. Let's start by finding out what's holding you back from your success and wealth.

Exposing the Programming

Be warned that the following exercise may trigger strong emotions inside you. That's OK. In fact, it's good. However, make sure to take adequate steps to protect yourself before beginning. At any point in this exercise, if your feelings start becoming too intense, or if you can't handle it for any reason whatsoever, stop reading immediately and seek whatever support you need. The purpose behind this exercise is to challenge you and release some pent-up emotions, not to push you over the edge. This is about growth, and must not be looked upon as any sort of therapy!

Most fears cannot withstand the test of careful scrutiny and analysis. When we expose our fears to the light of thoughtful examination they usually just evaporate.
Jack Canfield

Before you begin, take the time to create whatever supportive environment you need. For example, you may invite a trusted person to share this process with. If you do, then it is very important for the person listening to remain absolutely silent and supportive while you speak. This person must also agree to hold everything you say in absolute confidence, no matter what you say or how they feel about it. When you've created your supportive environment, you'll be ready to proceed.

Identify Your Beliefs

The *Exercises* section at the end of this chapter contains a number of statements that will help expose your core beliefs. Take all the time you need to fill each of these out. To do this, place a number next to each statement from 1 to 10. A one (1) means that you absolutely disagree with the statement, while a ten (10) means you absolutely agree. Most of your numbers will probably be somewhere in between. This is normal. Just don't be afraid to use either extreme if it applies to you.

You can change your beliefs so they empower your dreams and desires. Create a strong belief in yourself and what you want.
Marcia Wieder

After completing the statements, write down your feelings towards money, success, happiness, fulfillment, and dharma. Answer every question thoroughly and honestly. This may be hard to do because it may stir up long-dormant emotions. You may be moved to tears or anger, or may experience feelings of guilt, inadequacy, or just about anything else. It is critically important that you allow these emotions to flow through you, and for you to allow yourself to feel whatever it is you feel. It's OK. You need to release these bottled up emotions that have been poisoning you for so many years. Just remember to stop and seek appropriate help if at any time you feel yourself approaching a place you can't handle without additional support. Seriously. Don't let yourself go over the edge.

When you're finished, stop talking and/or writing. Do not under any circumstances discuss or analyze what just happened. If this is your story, you may be tempted to clarify or water down details. Don't! This is your brain throwing up obstacles, and is exactly what I mean when I say that the most growth occurs during the periods of greatest discomfort.

If you're the person listening, you may be dying to hear more. This is perfectly natural. Keep in mind that had your partner wanted you to know a detail, she or he would have told you. As the partner, you are the sounding board; you are not a judge, counselor, advisor, or questioner. Your sole job is to be the sponge that silently absorbs whatever you are about to hear. Remember that this is not intended as therapy. All you are doing is simply sharing and getting things off your chest.

If you're the person speaking, share every belief that you rated at 6, 7, or higher with your partner. Simply tell her or him the belief, and your thoughts and feelings around that belief.

Identify the Source

Whenever a thing is done for the first time, it releases a little demon.
Dave Sim

Your next step is to list the sources of your beliefs. Do you remember reading or hearing anything specific about money or success? If so, write it down. How did those people closest to you act around the topics of money and success? What specific life experiences affected your beliefs? How old were you when these experiences occurred? Write everything down. Here again, you may feel powerful emotions. That's OK. Just remember to stop and seek additional support if there is any risk that you can't handle what you're experiencing. If you want someone with you, that's perfectly all right, so long as you follow the same rules of silent support and confidentiality we just discussed. If you have a partner, go through each belief rated 6, 7, or higher. For each belief, restate the belief itself and share what you saw, heard, and/or experienced that might have formed or contributed to this belief.

As you go through this exercise, understand that what you're doing is exposing the reasons why you are not getting what you want in your life and why you're struggling without seeing the results you want and deserve.

Revealing the Ugliness

When you're ready, take a look at what you've written. Does this just about sum up everything that's been happening to you? Have you identified one or more endless cycles of ups and downs in your life? Is this all somewhat depressing or very depressing? You may feel like a total failure at this point. That's perfectly normal.

Life is raw material. We are artisans. We can sculpt our existence into something beautiful, or debase it into ugliness. It's in our hands.
Cathy Better

What I am about to say may sound heartless at first, but that is not my intention. On the contrary, by owning the truth I am about to tell you, you will be much further along your path to becoming an enlightened savage. I again repeat my earlier warnings to seek additional emotional support if necessary.

Remember this: Your reactions to the visual, modeling, and event learning you received from the moment of conception on formed your core beliefs, which unerringly led you to this point. You, and you alone, are absolutely, 100% responsible for your present situation. Did someone rip you off, use you, or otherwise take advantage of you? Guess what: You put yourself in a position for that to happen. You allowed it to happen, either through action or inaction. Of the two, inaction is far more common.

Coincidence? No such animal. Deepak Chopra talks of energy and the connectedness of all things. Your beliefs are a form of energy, and the Universe is a closed system. Chopra's theory therefore makes perfect sense. It also dovetails well with the standard Copenhagen interpretation of quantum physics that puts the observer squarely in the driver's seat of her or his own personal reality. Also, consider that the Universe literally started as a near-singularity, an extremely tiny dot many millions of times smaller than the period at the end of this sentence. At that moment, everything in the Universe was in a state of oneness. All of this matter and energy dispersed during the Big Bang and inflationary period that followed it, but everything still retains traces of having all been one and the same. In quantum physics, this unified state is referred to as *correlation* or *entanglement.* Entanglement exists at many levels, most of them too weak to be practical, but it still exists.

The possibility of stepping into a higher plane is quite real for everyone. It requires no force or effort or sacrifice. It involves little more than changing our ideas about what is normal.
Deepak Chopra

What we dismiss as coincidence is nothing more than a result of the inner lenses through which we view the Universe filtering what we pay attention to; this is similar to—but different from—what Deepak refers to as "synchro-destiny". It's a glorified way of saying that what goes around comes around. The Bible refers to this as reaping what one sows. Buddhists call it karma. Plant an apple seed, get an apple tree. It's that simple.

But back to you. As I just said, you have no one to blame or thank for your present condition but yourself. Are you a victim? If so, what is a victim but someone who blames their current problems on past events and/or other people? The real travesty is that society not only buys into this farce, we almost insist on it! How else do murderers get off lightly because they supposedly ate too many snack cakes (the infamous "Twinkie defense") or were abused as children? Your life events shaped your beliefs and may have been absolutely horrific. So what? The causes behind any of your effects are the causes behind all of your effects. You let those causes rule you, and now here you are, lacking the success you want but having all the success you deserve!

Again, you have no one to blame or thank but yourself. You own your life. Look in a mirror. You. Your beliefs. Your thoughts. Your emotional addictions. Your actions. Your results. In other words, you are seeing the effects of your inner self *realized* (made real) in your life to this point. To put it quite simply, you and you alone caused everything. There is no exception, and owning your responsibility is what allowed you to get to the point where you picked up this book. It's all you.

Revealing the Beauty

Give me beauty in the inward soul; may the outward and the inward man be at one.
Socrates

At this point, you may be wondering why the heck you just spent good money on a book whose author dares to insult you by rubbing your nose in your own failure. Your brain may be begging you to set this book down and never open it again. I understand and you know what? Thanks for the step up! Don't give in, because right now I really need you to hang on tight and stay with me.

Why? Why did I belabor the point that you brought this all down on your own head? Why harp on the fact that you followed your programming to the letter? And for Pete's sake, why on Earth would I make you feel like the failure you are if I'm trying to help you?

Do you want to know?

Are you ready to know?

Take a deep breath, because...

There is no such thing as failure!

I'll say it again: There is absolutely no such thing as failure. And if there is no such thing as failure, then you must be a total and complete success. But how can that possibly be? If you were a success, you'd have everything you wanted and then some. Something seems to not quite be adding up here.

Hello! That's your brain trying to shield you from the frightening truth: You, dear reader, formed your beliefs and conclusions and became addicted to the resulting emotions before you had the faintest ability to know what the heck was going on. You followed your core beliefs with unwavering loyalty and realized the personal reality specified by those beliefs. Remember the example we used of the ten computers running different programs? You have fulfilled your core programming to the letter. Believe me, this is a good thing!

Failure will never overtake me if my determination to succeed is strong enough.
Og Mandino

Do you understand? You did an utterly perfect job of following your beliefs, which became real themselves in your personal reality, which reinforced your beliefs, which built up the walls around your comfort zone a little higher, and around and around you've gone through the same cycle over and over again. That can only mean that you are a success!

Still not making sense? That's OK. Here's the final piece of the puzzle: You have already demonstrated that you follow your existing programming precisely as written. It therefore stands to reason that you could follow any other programming just as perfectly. Ponder that: What if your parents had treated money as a source of joy? What if I had understood why I couldn't have my microscope? What if any of the things you wrote in your workbook had been different? Think your life would have turned out a little differently? I know for a fact that mine would have.

My friend, the fact that you don't have all the success, wealth, joy, and fulfillment in your life in all probability means that you're perfectly healthy. Your brain is working properly by executing its programming to the letter. Like the computers in Chapter 4, the fact that your brain has perfectly executed flawed programming means that it can execute positive programming just as easily. Let me say this again: All of your trials, tribulations, and frustrations mean that everything you want in life is within your grasp! You can do it... and I'm going to show you how.

Questions

Please take all the time you need to complete this very important exercise by giving each of the following statements a score from 1 (totally disagree) to 10 (totally disagree). If you would like to invite a partner to help you, by all means do so. Remember that the listener must remain silent while you speak and must not question or judge what you say. If you're the person sharing, let it all out.

- **Money is inherently evil. ______**
- **I define wealth as money. _____**
- **Becoming wealthy will make me less human. _____**
- **Wealthy people are dishonest. _____**
- **Getting what I want in life takes struggle. _____**
- **I don't deserve to have what I want. _____**
- **Getting what I want means that others must do without. _____**
- **Happiness requires luck. _____**
- **I am not on this planet for a purpose. _____**
- **My life is not fulfilling any purpose. _____**
- **People who do whatever they want are selfish. _____**
- **It is wrong to put myself first. _____**
- **I am not good at managing money. _____**

- I am not good at managing my life. ______
- I live to serve others instead of myself. _____
- Chances are I'll never be happy. _____
- Chances are I'll never be wealthy. _____
- Living my desired life is too much hassle. _____
- I don't deserve to be happy. _____
- Happiness and wealth aren't for me. _____
- Those close to me won't support my changes. _____
- Living my dharma means exploiting others. _____
- People won't like it if I get rich or happy. _____
- People will want things from me if I'm rich. _____
- Getting what I want requires greed. _____
- If I strive for what I want and don't get it, I'll be a failure. _____
- All I need is a break to get what I want. _____
- This isn't the right time for me to start changing my life. _____
- I don't really want to be happy. _____
- I don't really want to live on purpose. _____
- I don't really want to be wealthy. _____
- Wealth and happiness aren't important. _____
- Having what I want will cause problems. _____

- **I'll never be wealthy doing what I love. _____**
- **It's hard to get ahead these days. _____**
- **I'm not smart enough to live my dream life. _____**
- **I'm too young to live for myself. _____**
- **It's too late for me to live for myself. _____**
- **I don't like selling or promoting myself. _____**
- **I wish I didn't have to worry about all this. _____**
- **I don't enjoy making time for myself. _____**
- **I find it difficult to relax and have fun. _____**
- **I need a steady job to be secure. _____**
- **If you're born into one lifestyle, changing it is extremely difficult. _____**
- **If I strive for what I want, I may lose what little I have. _____**
- **Happy people aren't wealthy. _____**
- **Wealthy people aren't happy. _____**
- **Happy people are corrupt or irresponsible. _____**
- **Wealthy people are corrupt or irresponsible. _____**
- **My happiness depends on getting the approval of others. _____**
- **People won't approve of my efforts to change my life. _____**
- **I don't know who to go to for support. _____**

- **I always put other people's wants and needs ahead of my own. _____**

- **I don't have enough time to spend with my family and friends. _____**

- **I have few close friends. ______**

- **I don't meet people or make friends easily. _____**

- **Becoming an enlightened savage is just another pipe dream. _____**

Feel free to insert more statements and score each one from 1 to 10 just as you did above.

- ____ _____
- ____ _____
- ____ _____
- ____ _____
- ____ _____
- ____ _____
- ____ _____
- ____ _____
- ____ _____
- ____ _____
- ____ _____
- ____ _____
- ____ _____
- ____ _____

Chapter 7

Visualizing Your Goals

What does success mean to you? The fact that you purchased *The Enlightened Savage* means that you have expectations and desires that aren't being met. For convenience, we call these expectations *goals* and their achievement *success.* The first step in achieving success is to clearly define your goals. In other words, what's the object of the game you're about to play? Seeing yourself having everything you've ever wanted is the first step to actually getting it.

Your Dreams are Real

Dreams come true; without that possibility, nature would not incite us to have them.
John Updike

Remember the process of realization? If you believe it, you will see it (within limits). I have three specific reasons for guiding you through this experience:

- Most people's brains operate in a so-called beta state with lots of high-frequency waves. Beta waves are associated with stress. Stress closes our minds, because it triggers the fight or flight response designed to cope with what? You guessed it: predators. Relaxing the brain triggers an alpha state that consists of wide, slow waves. Brains in an alpha state are at peace and can unleash their full creative potential.

- Visualizing your goals really is the first step to making them real. It's one thing to say you want a house, quite another to actually imagine the exact house you want.
- Brain studies can show which parts of the brain are active at any given time. Did you know that your brain's activity patterns are identical when processing both real and imagined information? At some level, your brain literally can't tell the difference between reality and fantasy. The potential implications are enormous. For example, are your dreams any less real than your waking life? I, for one, think they're just as real. But I digress.

In dreams begins responsibility.
William Butler Yeats

If your brain can't tell the difference between real and imagined information, and if some interpretations of quantum physics demand that the observer must impact her or his own observations—Do you see where I'm going with this? If not, then let me spell it out for you: You can theoretically build whatever personal reality you can imagine, because there is simply no difference.

But what about natural laws such as gravity? According to at least one scientist, natural laws only exist because we believe they do, which fits the "Law of Attraction" model perfectly. For example, if we believe with every fiber of our being that we can walk on water, then we can. The Bible tells the story of Peter walking on water because he believed he could. Realizing what he was doing, he got scared—and into the drink he went. This sounds great but ignores one very important fact: Each of us shares most of our realities with other people. I may believe that I can walk on water, but if you don't, then I can't. Shared reality is the lowest common denominator, without which the laws of physics that literally allow life to exist and things to make sense could not exist in any coherent fashion. More on this in Chapter 23.

What does this mean in the context of our discussion about goals? Again: If you can dream it, you can live it—if and only if you truly believe that you can, and then take proper action.

What else are we gonna live by if not dreams? We need to believe in something. What would really drive us crazy is to believe this reality we run into every day is all there is.
Jill Robinson

The Visualization Process

Take a moment to sit down in a comfortable chair. If you can put your feet up, please do. I want you as comfortable as possible. If you have a partner, by all means invite her or him to sit with you as we do this exercise. Remember that the partner's job is to simply listen without question, comment, or judgment.

A human being is part of a whole, called by us the Universe, a part limited in time and space. He experiences himself, his thoughts and feelings, as something separated from the rest—a kind of optical delusion of his consciousness.
Albert Einstein

Take a few moments to relax and clear your mind of all worries. Let your body melt into your chair. Place the tip of your tongue behind your upper teeth and hold it there. Take a deep breath. Hold it for a few seconds as you allow your concerns and worries to disappear. Then exhale slowly, feeling a wave of relaxation wash over you from head to toe. Let's do this again. Breathe in, hold, and exhale. And finally, let's repeat the deep breath, hold, and exhalation a third time.

With each exhalation, feel yourself relaxing further and further. Being calm and relaxed allows your mind to open, and to see and feel freely.

Keep the tip of your tongue against your top teeth. Close your eyes and take another deep breath. Hold it. As you're holding your breath, picture yourself doing, being, and having everything you want. What does that image look like? Do you want to travel? Where? What does your house look like? What color is it? Is it wood, stucco, or brick? Is it on a hill or in a valley? What's the climate like? What does each room look like? Fix this vision in your mind and then exhale, allowing what you see to wash over and surround you from head to toe, immersing you in your perfect ideal life.

Breathe normally and continue exploring this perfect world you created for yourself. Keep the tip of your tongue against your upper teeth and with each exhalation feel the relaxation, and savor the increasing vividness of your vision. What do you see? What do you hear? What do you feel, smell, and taste? Engage as many of your senses as possible, because the more senses you engage, the more real your vision will become.

Imagination is the voice of daring. If there is anything Godlike about God, it is that. He dared to imagine everything.
Henry Miller

While knowing what material things you want in your life is essential, it's only a tiny fraction of all there is to be and do. What do you want to learn? What do you want to accomplish? What do you want to do for others? How do you want to relate to others? How will you give back to this bountiful world you live in? Becoming an enlightened savage means examining and actively creating all parts of your life without exception. Keep your eyes closed, stay relaxed, and keep breathing in gentle, even breaths.

Take all the time you need to imagine everything you want. Material possessions, relationships, learning, spirituality, religion, travel, loved ones, whatever you want. If you want to go back to school, try imagining yourself on graduation day, holding your new diploma with honors.

Want to hike the Inca Trail in Peru? Picture yourself cresting the last hill, your lungs straining in the high altitude, and seeing the lost city of Macchu Picchu unfolding before you. That was my dream. I wanted to be looking down on Macchu Picchu at daybreak on my 40th birthday. And you know what? I realized that dream. I stood at a temple overlooking the city surrounded by mountains that looked like saw teeth with vertical cliffs and ragged tops. The sun was behind the mountains in the east and I could see the rays moving down toward me.

[There is a] no-win decision that innumerable people make, or try to make, or try to put off making daily: Whether to give up the job, the place, the people, the future one holds dear, denying one's own mental capacities, independence, and desires (what are left of them, what one remembers of them) just to get away.
Claudia Brodsky Lacour

If that was all, then I would achieved a life dream and been ecstatic. But that was only a small part of the magic of that day. My partner Jennifer met me in Aguas Calientes, the town at the foot of Macchu Picchu, the day before my birthday and went up to the temple with me. As the sun rose, I asked her to marry me. The moment she said yes, the sun rose over the peaks and bathed us in light. I don't have the words to describe the beauty of that moment. Being with her is unlike anything I've ever experienced. I do occasionally regret how my first marriage turned out, but nobody involved has any complaints about how it all turned out, not even my ex-wife.

If you have a partner with you, describe what you are seeing, hearing, feeling, smelling, tasting, thinking, and feeling as you explore your world. Partners, if you like, you may take notes to help recall the experience later.

Take all the time you need to experience every nuance of appearance, texture, temperature, smell, heat, taste, sound, feeling, thinking, and being. If you find yourself coming up with anything business-related, tune it out! Your business or job must contribute to your life, not the other way around. Don't worry about how you're going to pay for it all or what you'll need to get what you want. Just concentrate on the results.

For the duration of this vision, you have achieved your perfect life, and done so with style and grace. There is no right or wrong, no judgment at all. If it matters to you, then it's right. Your goals are yours. You have every right to every one of them. And as they say, to thine own self be true. If not you, then who else?

Get out of the blocks, run your race, stay relaxed. If you run your race, you'll win. Channel your energy. Focus.
Carol Lewis

Take a few more moments to explore this world you created for yourself. Finally, take everything you sensed, thought, and felt during your vision and pack it into a tiny glowing ball of energy. Then grasp that ball between your thumb and index finger to store it. Whenever you touch your thumb and index finger together, you will be transported right back to your vision.

When you're ready, lay the vision aside. Before you do, make everything you're sensing extra vivid. Commit it to memory for perfect recall whenever you need it. Then slowly let the vision fade and give yourself a pat on the back. You know what you want. And you'll get it.

So how was it? What did you see, hear, feel, smell, and taste? What did you think as you pictured yourself in your ideal life? Most importantly, how did that experience feel while you were in it? It is very important that you hold on to this feeling without qualifying, quantifying, or analyzing it.

Questions

The more detail you go into, the better. These questions will serve as prompts to help get you started if you're stuck, but don't stop there. Remember, the more, the better! Write down all of your answers. Make absolutely sure to express your goals as destinations, and don't worry about the dollar amounts. This exercise is designed to paint a broad picture of what you want from your life and what it will take to get there. Having defined where you want to be, the next step will be to define a map for how to get there.

- **Describe your ideal life in as much detail as possible, being sure to cover things like what you would like to see, hear, feel, smell, and taste. Include your home, business, loved ones, friends, family, location, travel, everything. If you want it, list it.**

- **How would you earn enough money to obtain and maintain this lifestyle?**

- **What other thoughts occur to you as you complete this exercise?**

Chapter 8

The Brass Rings

You've got to think high to rise. You've got to be sure of yourself before you can ever win a prize. Life's battles don't always go to the stronger or faster man; but soon or late the man who wins is the one who thinks he can.
Walter D. Wintle

In this chapter, you are going to take the results of the visualization exercise you just completed and begin mapping out how you will go about achieving them. Don't worry about the details for now. You'll create a detailed goal matrix and action plan later. For now, I just want you to have a glimpse of the road ahead.

Begin by reading over your notes from the previous exercise. These notes are your descriptions of your ideal life and are the entire reason why you purchased this book. The desired reality behind these notes is the brass ring, the reward you get for releasing all of the negative core beliefs that have been preventing you from realizing (making real) the reality you want—and deserve.

Types of Goals

At 50, everyone has the face he deserves.
George Orwell

As you review your notes, you'll start to discern specific goals, each of which will contribute to the vision you have for your ideal life. Write each goal down as it comes to you. You'll soon notice that you can fit each goal into one of three loose categories: Internal, external, and commercial.

Internal Goals

Internal goals don't involve anyone else because they are strictly personal. Achieving these goals might help you with other people; however, you are the main focus. Internal goals are therefore the most important goals of all. They are what you want to learn, what you want to experience, and who you want to be. Using the notes you took in the last chapter as a guideline, list some internal goals. Remember to only list those goals that are all about you. We'll talk about the others in just a few minutes. Don't worry about how you'll get there. We'll get to that a little later. Once you list your internal goals, give each a score from 1-10, with 1 being nice to do and 10 being absolutely critical life goals, things you positively must accomplish (or at least put concerted effort towards accomplishing) in this lifetime.

External Goals

External goals define both how you want to interact with others and what you want from these other people in terms of recognition, love, etc. Like internal goals, they have nothing to do with money. They merely center on your interactions with others. Take another look at your notes from the previous chapter and write out your external goals, then give each an importance score of 1-10. External goals are very important, but less so than internal goals. Ultimately, you are all you have. You must therefore take care of yourself first.

> *If you think you are beaten, you are; if you think you dare not, you don't. If you'd like to win, but think you can't, it's almost a cinch you won't. If you think you'll lose, you're lost, for out in the world we find that success begins with a fellow's will; it's all in the state of mind.*
> Walter D. Wintle

You may notice that some goals overlap. For example, you might want to feel confident about speaking in front of others. This could tie into an internal goal of raising self-esteem.

Take a few moments to see if there are any goals that have internal and external components. If so, break them up accordingly.

Commercial Goals

How much money do you want or need in order to achieve and maintain each of your listed internal and external goals? You may have listed a commercial goal that doesn't seem to tie into an internal or external goal. If you come across this situation, take a good hard look at that commercial goal and ask yourself why you want it. Remember that money is nothing more than a tool for accomplishing other goals.

Rank your commercial goals in order from 1-10 using the same scale as before. Remember that your commercial goals exist to help you achieve your internal and external goals. If you have one or more internal goals with scores of 6, 7, or higher, then those are your life's most important goals, because when you boil everything down, you're all you have.

Goals vs. Beliefs

In an earlier exercise, you listed your core beliefs and where they came from in terms of what you saw, heard, and experienced. Remember that your core programming exists because humans are prey animals and your brain accepted negative programming as survival instructions because it had no choice. Everything you've felt, thought, and done since then has reinforced that initial programming. So what if you didn't have this programming? What could you do and be if set free?

Whatever that is, the time to start accomplishing it is right now. Look at your listed goals, being sure to pay special attention to those ranked 6, 7, or higher. What's stopping you? Your negative core beliefs. How long will these beliefs last? For the rest of your life, or until you decide to change them, whichever comes first. When should you change them? Now. Not tomorrow. Not next week. Not when you get to it. Now.

You Don't Have Forever

Time rushes by and yet time is frozen. Funny how we get so exact about time at the end of life and at its beginning. She died at 6:08 or 3:46, we say, or the baby was born at 4:02. But in between, we slosh through huge swatches of time—weeks, months, years, decades even.
Helen Prejean

As Stephen King wrote in *The Shawshank Redemption*, get busy living or get busy dying, my friends. Every moment that passes where you are not actively seeking to accomplish your goals is one moment that is gone forever. The goals you listed are what you want to see, do, accomplish, and be. They are what you want to look back on when you reach the end of your time on Earth. And that is the one thing that unites all life: One day it is going to end as we know it. We are going to die.

As those last few minutes tick away and you look back on your life, too far gone to do or undo anything, what is it that you want to see? What do you want your legacy to be? This is an extremely important question, because your time is limited. You don't have all the time in the world. In the time it has taken you to read the last few sentences, another minute has slipped through all of our fingers. Another minute is gone forever and we can never have it back. It is another minute closer to the end of our lives.

If the Judeo-Christian-Islamic model is correct, then we will each face judgment for everything we have done in our lives by commission or omission. What will you say when God asks you how you lived your life and what you did with the time He gave you? If the Buddhist model is correct, then every moment wasted is a moment of negative karma that will keep you stuck in the cycle of birth and rebirth that keeps you from achieving divine union with Brahman that much longer. I happen to believe in a model that falls somewhere in between for reasons that I discuss in *The Divine Savage.* The science and philosophy behind my model is absolutely sound, but what if I'm wrong? What if everyone who believes in some form of afterlife is wrong and death really is the abyss of nonexistence? Any way you slice it, none of us has a second to lose. The sound of the clock on the wall is the sound of life ticking away.

Does that idea scare you? It used to petrify me. But now I see it as an incredibly enlightening thought that no longer frightens me. Why? Because I've found my life's mission and am fulfilling it.

Imagine Success

Let's do another exercise. If you have a partner, invite them to sit with you. When you are ready, have a seat, get comfortable, and relax. Take in a deep breath. Hold it as long as you can, then let it out. As you let this breath go, feel a wave of relaxation coming down your body from head to toe. Do this a second time. In, hold, and exhale, feeling the relaxation wash over you. Breathe in, hold, and release a third time.

You or your partner will now read each goal rated a 6, 7, or higher aloud. After you hear each goal, repeat it back to your partner as if you had already accomplished it. For example, if your partner says "I want to become a concert pianist", you respond "I am a concert pianist." As you utter each statement, imagine yourself having accomplished that goal, then tell your

What is a television apparatus to man, who has only to shut his eyes to see the most inaccessible regions of the seen and the never seen, who has only to imagine in order to pierce through walls and cause all the planetary Baghdads of his dreams to rise from the dust.
Salvador Dali

partner how it feels. If this is an external goal, describe how the affected people feel as well. Touch your thumb and forefinger together to connect with the positive feelings you got in your previous visualizations. How does it feel to have accomplished this goal? Write down the emotion or emotions you connect with accomplishing this goal. If you have a partner, she or he can help you with this.

How should you go about accomplishing this goal, if you know? If you don't know, what can you do to learn more about this goal and what it takes to achieve it? Write down the steps you need to take to learn enough about this goal to be able to start realizing it.

Now we're going to repeat this exercise for every goal you rated as a 6, 7, or higher. Again, the process is as follows: You or your partner reads the stated goal aloud. You then respond with a positive statement as if you had already achieved that goal. Remember to let yourself feel each accomplishment.

Listen, if there's one sure-fire rule that I have learned in this business, it's that I don't know anything about human nature.
Francis Ford Coppola

Write down your emotional responses to the idea of achieving this goal, and briefly outline some steps to achieve it. Be sure to do this for each goal you rated as a 6, 7, or higher. And be sure to let yourself feel. Remember that the emotions are the source of everything, the root of the process of realization. The first step in achieving any goal is feeling it.

Your important goals, the sevens and higher, are the brass rings. These goals are why you want to become an enlightened savage and why you want to use your prey hardware and your prey instincts to your advantage. We can't rewire our brains into predator brains. What we can do is insert new programming into our prey brains that redefines what might get us killed and eaten. The idea is to reset your survival instincts to embrace gain and adventure instead of shrinking back from your goals and life's mission.

If the single largest thing separating humans from monkeys is, in fact, a mutated gene, then humans are arguably diseased monkeys. This disease reveals itself in unmatched intelligence that includes desire and purpose in a thin layer above the primitive brain that hides from dragons. On one hand, this is both a chronic and terminal disease that must be treated just like any other disease. On the other hand, this disease is what makes us human. It's also the only disease I know of where the treatment not only cures but improves us.

Knowing what your goals and priorities are gives you all the incentive you need to change your old programming and use your primal, prey instincts to their full advantage to obtain wealth and happiness; in other words, to become an enlightened savage. if you need more incentive, remember that your time on this planet is limited. Nobody has forever.

Put Yourself First

To say less of yourself than is true is stupidity, not modesty. To pay yourself less than you are worth is cowardice and pusillanimity.
Michel de Montaigne

Here's a word of caution: Society trains us to put other people ahead of ourselves. It might therefore be difficult to focus on your internal goals at first. Ignore that! You must live life on your terms. Don't hold back. If a goal really matters to you, then go for it. If anyone asks, tell them the truth: You can't help anyone else unless and until you help yourself above all others. Nobody can.

Yes, there is a fine line between being good to yourself and taking unfair advantage of others. However, consider this: I was a volunteer firefighter and emergency medical technician. Without that training, I would have been utterly unprepared to help people, and could have easily caused harm. You get the idea.

There is an old saying that if everyone tended their own little piece of the Earth, then this planet would be a paradise. I believe that. Most of us spend our lives caring for people, places, and things too far removed from our own lives to matter. The same lesson applies to life: If everyone took full responsibility for their own lives, my guess is that people would get along far better than we do.

As it stands, society seems to hold people responsible for everybody's life but their own. How else do you explain people getting sued because someone didn't bother looking where they were walking? We're not saying you should live your life in a vacuum. Far from it. In fact, mentioning other people leads us to external goals. Again, these are what you give others, or vice-versa.

I once read a poem posted in a bus:

Although I am but a speck of something small, I vow that when I leave it, the world will know that I was here.

I don't know who wrote them, but those 24 words became my driving force. My primary internal goal is to continue to develop the art of conscious observation and my conscious interrupt, which I'll talk about later in this book. My primary external goal is to use the *Savage* books and related products to positively influence people's lives and leave the world a better place than when I found it. What are your internal and external goals? What about your commercial goals? They exist only to fuel your internal and external goals.

Question: How well do your goals mesh with what your group (your family, friends, teachers, news media, etc.) taught you? Remember my friend who wanted to start a restaurant and was warned off by the people in her group who think that business is too hard because of their own struggles? Yes, you need to fit into a group because that's the way humans are designed. But you have the choice to fit into any group you want. How about finding and bonding with people who share your dreams and who know they can make real whatever personal reality they want?

Honor, riches, marriage blessing, long continuance, and increasing, hourly joys be still upon you! Juno sings her blessings on you.
William Shakespeare

Remember: If you want to be rich, learn from a rich person. If you want to fly, learn from an experienced pilot. If you want to realize your ideal life, learn from people who are doing that in their own lives. Build the social structure you need to support your goals, not the other way around!

Never forget that your commercial goals only exist to help you achieve your internal and external goals. Your business must exist to serve you and your needs, not the other way around.

Most people confuse jobs with businesses and treat them the same. At a job, your goal is to serve the company's needs, not yours. Most people take that mentality to their own businesses with predictable, tragic, and unnecessary results. Don't fall into this trap. Take care of yourself first!

Questions

What are your life goals? How important are they to you? They can be trivial, vital, or anywhere in between; there are no wrong answers. *The Enlightened Savage* exists to help you achieve your internal, external, and commercial goals.

- **List your internal goals and score each one from 1 (not important) to 10 (extremely important).**

- **List your external goals and score each one from 1 (not important) to 10 (extremely important).**

- **List your commercial goals and score each one from 1 (not important) to 10 (extremely important).**

- **How are the beliefs you identified in the previous chapter helping or hindering your progress towards achieving your goals?**

- **Are you willing to change your unsupportive beliefs in order to begin accomplishing your goals?**

- **What does achieving your most important goals look, feel, sound, smell, and/or taste like? Describe in as much detail as possible.**

Chapter 9

Dharma

Nature arms each man with some faculty which enables him to do easily some feat impossible to any other, and thus makes him necessary to society.
Ralph Waldo Emerson

The probabilistic nature of quantum physics means that all we can know are probabilities until we make an observation. Even then, the Heisenberg Uncertainty Principle limits how much we can know about any observation. Einstein proved that matter and energy are absolutely interchangeable, that $e=mc^2$. His theories of relativity even showed that space and time are inextricably linked. Many experiments have proven that the line between matter and energy is fuzzy at best. Even space and time seem to be illusions. As astrophysicist James Jeans said, "the Universe begins to look more like a great thought than like a great machine." Randomness and uncertainty appear to rule the quantum world, but the Universe as a whole bears all the hallmarks of intentionality and purpose.

I delve into all of this in *The Divine Savage.* Meanwhile, the important point now is that you are on this planet for a reason. You are meant to contribute something to humankind and to the Universe. You have a mission, whether you choose to accept it or not. The sooner you find your mission, the more successful you will be in this lifetime. Combining your mission with changing your beliefs delivers the magic 1-2 punch on your road to success. Can you achieve some success by doing one or the other? Sure. But that success will be limited because of the imbalance between your beliefs and your mission.

Knowing your mission reinforces your new beliefs, and your new beliefs reinforce your sense of purpose. So far, we've spent some time talking about your existing beliefs and goals. Now let's explore a few ways in which you can discover your purpose in life. You might think this a bit redundant because just you've finished visualizing and writing down your goals. It isn't. Knowing your goals is critical because your time here is limited, and nobody can say for sure what comes after this life... if anything. We may think we know—I for one am virtually certain—but in the end there is only one way to find out. I don't know about you, but I'm certainly in no rush!

Our goals are what we want to do and be. But are they what we need to do and be? Chances are very good that your stated goals are at least part of your greater mission. If so, then this chapter will reinforce what you already know. If not, isn't it best to find out now?

About Dharma

Dr. Wayne Dyer, creator of the *Power with Intention* seminars, asks a very simple question: Have you lived ten thousand days, or one day ten thousand times? If you're like most people, you've been stuck in the same rut for as long as you can recall because of your self-destructive beliefs. We've already identified those beliefs and their sources. We've also talked about your goals. The next step is to find out just what your mission or dharma is in order to ensure balance. The word *dharma* means "The essential function or nature of a thing." What is your essential function or nature and how do you go about finding it?

Belonging to oneself—the whole essence of life lies in that.
Ivan Turgenev

Dharma and Health

We're all familiar with physical and mental health problems. But what about not knowing your life's mission? Is that a health problem? The word *disease* means a lack of ease. If you're in a rut or wandering aimlessly, then you aren't experiencing a lot of ease because you are not living your dharma.

Any important disease whose causality is murky, and for which treatment is ineffectual, tends to be awash in significance.
Susan Sontag

Here's a very small example: While writing the first edition of this book, I found myself struggling with this chapter and needed additional information. I waffled, using my writer's block as an excuse to stall my progress. Out of nowhere, I came down with a nasty cold. After a couple of days of suffering, it occurred on me that milking my situation might have something to do with my condition. I threw myself back into my work, and my nose cleared up. What happened next? My throat got so sore that I couldn't speak for several days. This effectively cleared my calendar and gave me lots of time to read. I listened to those symptoms, did my research, and my symptoms went away. Think about those symptoms: Blockage signified blockage, and inability to communicate with others opened up free up time for writing. Makes you wonder, doesn't it? Especially if you share my conviction that there is ultimately no such thing as randomness, accident, or coincidence... only purpose.

Remember the ETEAR process of realization: We become addicted to our emotions and create toxins when we experience negative emotions. These toxins can even accelerate the aging process. That's right: The worse you feel, the less time you may have to live! By contrast, I've read many examples of dramatic healing that took place when people listened to—and acted on—their dharma. Modern society seeks to eliminate pain to achieve better living through pharmacology, as a friend once put it. I used to agree with that mentality until I began seeing my aches and pains as messages and guides that appeared because I had ignored earlier, more subtle warnings. I'll say it again: Dharma is inescapable. If dharma is a health issue, it follows that obeying one's calling must be medicinal.

Using this definition, disease of the dharma is very serious. Symptoms can include depression, anxiety, fatigue, self-doubt, fear, troubled relationships, struggle, and even physical illness. Therefore, not knowing and/or not following your dharma is indeed a health problem. The cure is very simple: As psychologist James Hillman said, you must give up the life you have to get to the life that's waiting for you.

Every human being has, like Socrates, an attendant spirit; and wise are they who obey its signals. If it does not always tell us what to do, it always cautions us what not to do.
Linda M. Child

Signals

Every noise you hear is caused by vibrations carried through the air. You can get a good idea of how sound works by dropping a stone into a still pond and watching the ripples. (Those ripples are also a god approximation for what matter looks like in the absence of observation.) Air is the medium that carries the sound to your ear. Sound cannot be transmitted without a medium. For example, if you were in space, you would hear nothing. You wouldn't even a speaker cranking out heavy metal tunes at full volume right beside you.

Air carries lots of information including sound, broadcast signals, scents, pollen, and pheromones, of which humans can only perceive a tiny fraction. Pollen might excite your allergies, but it literally means life for a plant species. You might not be able to smell a mouse hiding in the bushes, but a cat sure can. The fact you can't perceive all this information doesn't mean it isn't there. You're just attuned to a very small part of it.

Like the air around you, your life is filled with signals pointing you in the right direction. Most people think that the physical world they inhabit is the cause of their personal reality and not the result. We impose so many conditions on the miracles we receive every day that their messages and guidance are hopelessly lost. If we do get the message, our brains concoct every imaginable excuse to keep us in our burrows. Why? Because of the negative core beliefs carried by a prey brain that interprets any deviation from your routine, no matter how bad it is, as mortal danger. This is the same fear of getting killed and eaten that we've been talking about all along.

I don't know Who—or what—put the question, I don't know when it was put. I don't even remember answering. But at some moment I did answer Yes to Someone—or Something—and from that hour I was certain that existence is meaningful and that, therefore, my life, in self-surrender, had a goal.
Dag Hammarskjöld

Our ability to change is what we fear the most because that change comes from the act of leaving our burrow at noon, when we already know that hungry eagles are circling above. Marianne Williamson says that is not our darkness we fear but our light, not that we can't but that we can. this makes perfect sense. The gopher in the burrow is not afraid of his ability to remain safely underground in the darkness. He is afraid that he is perfectly capable of emerging into the daylight and becoming lunch. That fear keeps us safely rooted in our unfulfilled lives because our lives are safe and manageable on some level, no matter how bad they might be. Tuning into the signals and finding your dharma means leaving your burrow and embarking on a journey that may not always seem to make sense, but that still draws you inexorably forward.

Finding Your Dharma

Nature speaks in symbols and in signs.
John Bartlett

How do you find your dharma? Listen! Listen to everything around you. Be open to your daily blessing without preconditions. The worse (or better) something looks at first, the greater the opportunity buried deep inside. There is nothing intrinsically god or bad about anything in life. Positivity and negativity are labels we ourselves apply. The key is to apply those labels consciously, as we'll discuss later on. For now, the important thing to keep in mind is that the stronger the potential emotional charge, the greater the potential opportunity for you. In case I haven't made this clear enough, this applies equally to any event in life, whether we label it positive or negative. Be willing to follow your dharma when it comes and to seize opportunity when it knocks. Remove all preconceptions you might have about who you are and who you are meant to be. Most of all, when the call comes, act !

Surrender to Your Power

We're very accustomed to trying to impose our will on ourselves and our worlds. By contrast, finding your life's mission is more about asking questions than about making hard decisions. This is why you need to stop telling and start asking questions—and listening to the answers, no matter how unexpected they may be. Fulfilling your dharma means devoting your life to seeking the answer to your question or questions.

In the last chapter, I mentioned the importance of surrounding yourself with like-minded people. The moment you begin your journey of self-discovery, you will find yourself drawn to others who share your path and who can help guide and advise you. Never forget that there is no such animal as coincidence!

Clues are Everywhere

The fundament upon which all our knowledge and learning rests is the inexplicable.
Arthur Schopenhauer

How do you find and define the question? In his excellent book, *Callings: Finding and Following An Authentic Life*, Gregg Levoy lists some of the many ways our dharma comes to us. They include recurring dreams or whatever pursues you during dreams. Pay special attention to lucid dreams. I suggest that you keep a dream journal and ponder a question, problem, or fantasy as you drift off to sleep. Dreams allow our minds to explore any imaginable subject without any constraints. Don't forget that part of your brain can't tell the difference between the waking world and dreams. Harness that!

How about physical symptoms that are metaphors for issues in your life? What obstacles or challenges are you facing in life? Are they opportunities or curses? Remember that whatever you believe is real to you and forms your personal reality. If you decide to see them as gateways for growth, that's exactly what they'll be. Thanks for the step up! Ever have a song, conversation, or something else stuck in your mind? The poster I read on the bus as a child remains with me to this day.

I dismissed that call for many years and struggled because of it. The moment I decided to heed my dharma, my life began changing. My dharma is to make a positive difference in people's lives. If *The Enlightened Savage* does this for you, then my job is done.

Your vision will become clear only when you can look into your own heart… Who looks outside, dreams; who looks inside, awakes.
Carl Jung

Have you ever gotten strange urges out of the blue? In the Introduction to *The Divine Savage*, I talk about how I was watching TV one night and suddenly got the overpowering urge to read Genesis 1 in the Bible; a very odd thing, since I'm not religious in the traditional sense of the word. That incident was the straw that prompted me to begin trying to figure out the ultimate Truth, whatever that might be. My subsequent research led to the second edition of this book that you are now reading, *The Natural Savage*, *The Divine Savage*, *The Romantic Savage* (coming in 2013), and plans for four more *Savage* books. My studies have revealed a Universe full of purpose and intention, and we are all part of that grand scheme. Physicist Steven Weinberg could not have been more wrong when he declared that, "The more the Universe seems comprehensible, the more it also seems pointless."

I'm a member of Toastmasters, an international organization composed of local clubs that meet to practice and hone speaking skills. If you could stand up and deliver any message you wanted to a spellbound audience, what message would that be? What decisions are pending in your life right now? Here's a question: What do you fear the most? A speaking instructor once said that stage fright only happens to people with important things to say. Their words have a lot at stake, hence the fear. Likewise, your fears are windows to your life's mission.

My friend, all of the above are clues that can point you to your own dharma, and it is far from a complete list. As Gregg Levoy asks, what is right for you and where are you willing to be led? Balancing those questions and asking them again and again is the shortest path to finding your life's mission that I know of. Doing this literally focuses your brain to help give you more clues and insight. Some interpretations of quantum physics postulate that the observer influences—if not outright creates—the outcome. Seeking an outcome must therefore be part of the process of obtaining that outcome.

Two roads diverged in a wood, and I—I took the one less traveled by, and that has made all the difference.
Robert Frost

The Path Less Traveled

For all our strengths, humans are amazingly fragile creatures. Our fragility is our greatest strength, thanks to our ability to be changed and led in new directions.

The basic concept is the same as a vaccine which delivers a tiny dose of some awful disease in order to stimulate the body into producing antibodies that will ward off a full-fledged attack. Allowing that small injury prevents a far worse one from occurring. You can't avoid your dharma any more than you can avoid food and water. Dharma is the ultimate creditor. Avoid creditors and they just turn up the heat. (Trust me, I know this from personal experience!) This hurts and could even ruin your material life. Facing creditors head on reduces or eliminates the potential damage. Staying in your burrow at noon is the wide, easy path to destruction. Coming out is the narrow, steep, poison-oak-laden trail to fulfillment. I know from steep, narrow trails, because I love to hike as often as possible, including about 80 miles across an arm of the Peruvian Andes. I'm also highly sensitive to poison oak.

Sacrifice is nothing other than the production of sacred things.
Georges Bataille

Sacrifice

Human nature is to preserve the status quo until the pain of doing so exceeds the pain of change. Flood a gopher burrow and the gophers will emerge because the fear of definite drowning is stronger than the fear of potential predators. They might die if they leave their hole, but they will die if they remain.

This brings up the idea of sacrifice. Your dharma defines your entire reason for being. Following that path is both the greatest challenge you'll ever face and the greatest reward you'll ever get. You must therefore be willing to give up just about everything and everyone you hold dear in order to find everything and everyone that matters. Learning how to surrender what you have teaches you to receive what is coming. If you accept your daily blessings and believe in your own success, then what is coming will be far, far better than what you have. You will experience wealth, financial or otherwise, beyond your imagination.

Wealth means having everything you need to follow your life's mission. You might not get the fancy house and car. In fact, you might need to shed all of it as I had to do. Real wealth is being truthful to yourself and others. The saying "to thine own self be true" is the ultimate expression of wealth. Don't get me wrong: I am certainly not advocating a life of poverty or deprivation. Embrace your dharma and you will always have everything you need.

Decision is a risk rooted in the courage of being free.
Paul Tillich

You must be prepared to give your all. Half-baked efforts just won't cut it. Waiting for the perfect time to do something? Every moment you wait is a moment that's gone forever. There is no right time until you make the time right.

Sacrifice is the ultimate act of faith in yourself, your fellow people, and the Universe. Letting go and allowing yourself to be vulnerable is your ultimate strength. You will need that strength, for you will face every obstacle imaginable. Friends and family may abandon you. You may lose a job or have to sell—or even lose—your home. Following your life's mission is a never-ending series of tests that will have you either scurrying back to your former life or breaking through your obstacles. Are you sensing a theme in this chapter and across everything we've covered so far? I sure hope so!

Why are we doing this? Why are we seeking to find and fulfill a mission? Gregg Levoy explains it this way: Toss a stone into water and the water level will rise because the stone displaces the water. The difference might not be visible, but that doesn't make it any less real. At its simplest, dharma is the stone that lifts the water. Be that stone!

Ask, and it shall be given you; seek, and ye shall find; knock, and it shall be opened unto you.
Matthew 7:7

Seek and You Will Find

We all use tools to sense things beyond our direct perception. Your TV turns invisible energy into pictures and sound. These tools work because you want them to. You turn on the radio and listen to music. You open *The Enlightened Savage* and read the words I wrote. This book's effectiveness depends entirely on your reading, understanding, and heeding its words.

Same with your dharma. The more you listen for clues and guidance, the more you will receive. Practice your listening skills by raising your awareness. There are many ways to do this. Yoga, meditation, prayer, a dream journal, exercise, keeping a diary, conversations with trusted friends—Experiment until you find something that's right for you. I love hiking on steep trails. This strenuous workout lulls me into a calm state where I can hear myself think. Use whatever method works for you.

Do not seek death. Death will find you. But seek the road which makes death a fulfillment.
Dag Hammarskjöld

I strongly recommend that you choose something physical, because growing mind and body at once is a great way to open yourself to insight and inspiration—a very liberating experience. What you do doesn't really matter, as long as you enjoy it and it frees you from the rat race long enough to clear your head and tune yourself into what's happening around you. Some animals sit motionless for hours at a time. You can bet that they are acutely aware of their surroundings. Humans, by contrast, are so conditioned to be busy and "productive" that we often lose the ability to just be for a time.

When the answer to your constant asking comes, go for it with all the gusto you have, but never stop asking questions, because the last thing you want is to follow any path blindly. If your answer seems expedient, it's probably not right.

If you seek truth, you will not seek victory by dishonorable means, and if you find truth, you will become invincible.
Epictetus

If Answers Don't Come

Your dharma may or may not seem very clear to you, especially after the visualizations we did in the last two chapters. If not, that's OK. Just remain open and don't succumb to the desire to analyze everything to death. Analysis paralysis occurs when thinking replaces doing. Will you ever have all the information you need? No. That is where faith and intuition come in. Gregg Levoy suggests logging your hunches and seeing how often you're right. Give it a shot; the results might astonish you, because your personal reality is your result, not your cause. This process serves as the checks and balances that ensure that the new programming we're about to give you will work properly.

Passions are generally roused from great conflict.
Titus Livius

Embrace Conflict

Earlier, I suggested that following your dharma can lead to internal and external conflict. The paradox of following your calling is the seeming difficulty. But paradox exists everywhere in life. Watch as you move your arm and you'll notice that some muscles are expanding while others are contracting.

Your thoughts and feelings, facts and intuition, and belief and faith will pull you in every direction. To think that facing this head-on is the vaccine that prevents the disease! The more you can use this paradox to your advantage, the more powerful your results will be, because holding and processing conflicting thoughts in your head requires power. This power comes from openness and exploration, not a mad rush to some contrived resolution. In other words, the more you delve into these seeming contradictions, the stronger you'll become. As I said in Chapter 5, the shortest and most effective path is often plowing straight through an obstacle.

Shortcuts make for long delays.
J.R.R. Tolkien

No Shortcuts

The fastest path forward is the one without shortcuts. You want and need every one of your obstacles, because you can either let them fence you in or use them as stepping stones. Thanks for the step up! You must do whatever it takes to become the person you are meant to be.

While your dharma belongs to you, it isn't just about you. You will receive all kinds of external hints about whether you're on the right path. When I lost my marriage, income, and house in 2008 and 2009, I realized that it was a blessing because the Universe would no longer tolerate my aimless drifting and skewed priorities. I also realized that my experiences and the path I am taking to freedom can help others do the same in their lives. Writing the *Savage* books used to be a hobby. It is now my driving passion in life.

Several years ago, I realized that many authors don't know how to market their books and decided to create a series of videos and workbooks to help them implement and launch effective marketing plans. My friend Rose invested hundreds of hours helping me create those videos, processes that required several grueling videotaping sessions in hundred-degree weather, bumpy plane rides, long drives, and countless odd looks from passers by. Many times I wanted to quit, but Rose never gave up. At a time when I doubted my direction, her faith, strength, and support kept me going. Her belief in what we were doing led me to ask Jay Conrad Levinson, the father of Guerrilla Marketing, the man who changed how business does business, to work with me. The connections I have made since meeting Jay led to my finally getting my degree in psychology, being asked to become an executive in a ground-floor startup, and to meeting and working with renowned coach and trainer Jim Britt. All of this has led to you holding this book in your hands right now. I did not "attract" or "manifest" any of this. I took steps to make each of these things real.

One can never pay in gratitude; one can only pay "in kind" somewhere else in life.
Anne Morrow Lindbergh

When I finally opened myself to my dharma, the Universe responded by giving me sign after unmistakable sign indicating that I was moving in the right direction. If you follow your own dharma, you, too, will receive clear signs. Needless to say, I am eternally grateful to my friend Rose and to all those who are helping me on my way.

Payback

The preceding story brings up a question: Do you owe your guides and mentors for their efforts? Our society thinks in terms of transactions, of "Give me X and I'll give you Y." Thanks to quantum physics, however, science is finally beginning to understand in practical terms what priests, shamans, and mystics have been saying for thousands of years: Our universe is composed of inseparable parts. There is a deep purpose beneath the chaos and uncertainty of physics in which there is no such thing as randomness or coincidence. Events in the natural world can be compared to events in daily life and related to our rational and emotional minds. Newtonian cause and effect fails to explain most of what we experience, yet our experiences are all around us. As Gregg Levoy says, most of what we term *paranormal* (such as psychics, miracles, or ghosts) actually suggests that something—or someone—is standing behind the scenes controlling what's happening. To the religious person, this mysterious force or entity is God.

Given that, yes, I believe that you owe the entire Universe nothing less than your successful, ongoing achievement of your dharma, your life's mission. The best way for me to repay Rose is to succeed in the endeavors that she facilitated. That goes far beyond mere money.

Walk while ye have the night for morn, light breakfast bringer, morroweth whereon every past shall full fast sleep.
James Joyce

Moving Forward

Be open to following your dharma without limits. Remember that fulfilling your mission is all about serving your higher purpose. This may place you in a position of power and responsibility, but only if it's needed for your mission. Power never comes for power's sake.

Your prey mind uses fear to keep you in your comfort zone. You bought *The Enlightened Savage*, which tells me that you feel the need to change your comfort zone. That's a scary prospect, because your brain interprets any departure from its routing as risking getting killed and eaten.

What to do? Confront your fears! The worse the fear, the better the knowledge and wisdom you'll find lurking underneath. As Gregg Levoy says, the entrée called enlightenment comes with a side dish of holy terror. I had a terrible fear of death that I finally realized came from not fulfilling my dharma. I'm a private pilot. Common sense says that one must raise the nose to stop descending, but sometimes you actually have to lower the nose, literally pushing down to stop going down. Take your fears head on, and use all the unconventional tactics you can think of to work through them. That's what being an enlightened savage is all about. As for me, the moment I decided to start really living, my fears subsided.

Abraham Maslow spoke of shirking or avoiding your dharma as spiritual and emotional truancy. This truancy comes from evasion and self-destruction caused by negative core beliefs, primarily those that say that following dharma requires lots of energy. Well guess what: Avoiding dharma is far more costly.

Besides, as Marianne Williamson says, our deepest fear is not that we are inadequate, but that we are powerful beyond measure. It is our light, not our darkness, that most frightens us. It is the idea that we can physically walk out of our burrow at any time that petrifies us. Why? If you couldn't leave your burrow, then you'd have no reason to fear the hungry hawk circling above. The fact that you are indeed physically capable of leaving your burrow is what frightens you into staying put.

Remember the groups each of us belongs to? Each of the people in these groups has her or his own ideas about success and following dreams and dharma. These ideas follow the group norms; every group member can smell the moment someone starts moving to the beat of a different drum. And they won't sit idly by. Just as a wolf pack pounces on any member that steps out of line, so will your friends, family, coworkers, and others. Why? Because by taking action, you are upsetting the apple cart. This means that your efforts to find and follow your dharma may meet with opposition.

If you think you're outclassed, you are.
Walter D. Wintle

Where does this opposition come from? People will see you succeeding and may fear that it could happen to them, thus exposing their own negative core beliefs. Following dharma exposes friends and detractors. Rose and other friends invested in video equipment and helped me produce my first videos. You, too, will find help in unexpected places. And your detractors? Whenever someone says anything negative, take a few moments to objectively consider the source. What's the agenda behind their words? You may lose them as friends, associates, etc. But do you really want them around? My guess is no. You may need to find or build new groups and surround yourself with new people. That's perfectly OK! Remember, you are all you have. You owe it to yourself to live your life on your terms.

Above all, ask for help and guidance from any source and be prepared for the answers. They may not be the answers you want, but they will always be the answers you need.

We spent some time visualizing your goals in Chapter 7. The end of this chapter has exercises that will help you uncover your dharma. Your vision notes, goals, and scores are the jumping-off point. As you seek your dharma, you may find yourself adding or removing goals and changing their stated importance. That's totally OK. Remember, the whole point is to keep asking questions, not to seek answers.

So what if you keep asking but aren't getting answers? What if you're lost in the jungle with no idea how to proceed? This may happen for any number of reasons and is very common. If you're having rough going, try turning the problem around. Imagine yourself on your deathbed, about to discover what lies beyond the gate we must all traverse. It is too late for you to say or do anything. The end of this life is only seconds away. What will it take for that final moment of life as you know it to be a fulfillment? What will you need to have seen, done, been, experienced, learned, taken, given, etc. in order to die well? The fear of death is nothing more than the fear of a life not lived. So what kind of a life must you live in order to be able to leave this world in peace?

Somebody should tell us, right at the start of our lives, that we are dying. Then we might live life to the limit, every minute of every day. Do it! I say. Whatever you want to do, do it now!
Michael Landon

Be as specific as possible! Once you have a reasonably clear idea of what you want your final vision to look like, you can start to work back toward the present to decide what actions and which paths will help you best achieve your end. For life, you see, is simply preparing for death. Build the life that will let you die without fear and that will leave all those around you inspired to do the same. That is your dharma. That is everyone's dharma. How each of us accomplishes that one task is up to us. You can't win. You can't break even. You can't escape this truth.

Please don't be frightened or discouraged. Use the fact of your limited existence to get out there and start living the life you were meant to live in accordance with the Universe's grand design. The moment you start doing that, you will find any fears you may have about your own death easing. The more you build and live your ideal life, the less fear you will have. By getting busy living, you need not worry about dying. Finding and living your dharma is the key to peace, now and in any life that may come after this one. In fact, if the Buddhists are correct about karma and reincarnation, the trials and travails you are facing in this lifetime might well be caused by negative karma from past lifetimes. If this is true, then you owe it to your current and future selves to live your dharma!

I hope you're starting to see how all of the information we've given you is starting to fit together. In the end, how well you follow your dharma boils down to the reality you choose to build for yourself, a reality that is only limited by what you believe the limits to be (on a personal level) and the laws of physics (on the shared level)! The next step is to change your negative programming and then learn how to live consciously. We'll get to that next.

Questions

Please take all the time you need to answer the following questions, which are designed to get you thinking about your dharma. If you have not read Gregg Levoy's book *Callings*, order a copy today. It's an absolute must-read.

- **Do you believe you are on this planet for a purpose? If so, what? If not, why not?**

- **Describe your life's purpose in as much detail as possible.**

- **Do you suffer from any physical or mental ailments or diseases? If so, how do these problems serve as metaphors for what's happening in your life?**

- **Have you ever gotten sick or injured when confronted by an obstacle or challenge? Describe the circumstances, the ailment, and how the situation resolved itself.**

- **What other diseases of the dharma are you experiencing or have you experienced?**

- **Do you see these illnesses or injuries as trying to steer you down a different path?**

- **What obstacles and challenges are you facing in your life?**

- **Have you faced these obstacles before? If so, do you find them getting more serious over time?**

- **Describe odd impulses or urges you've felt. What do you think they were trying to tell you?**

- **Are you facing any major decisions?**

- **What odd things have you experienced that might serve as clues? Include things like songs coming on** at just the right time, meeting people as if by chance, etc.

- Do you feel as if your life is on the right path? Why or why not?

- Are you willing to give up everything you have in order to obtain everything you must have? If not, why not?

- Are you prepared to dedicate your entire life to living your dharma even if that requires large sacrifices?

- If you are unwilling or believe yourself unable to live your dharma, what regrets do you think you might have as you approach the end of this lifetime?

- What physical, mental, emotional, and spiritual exercises do you perform regularly to keep yourself in shape? Are these enough? If not, what else are you willing to do?

- Who has done you favors or gone out of their way for you? How can you reinvest their kindness in your own life and those of others?

- What is your worst fear? Does it have anything to do with your dharma?

- Do you have recurring dreams/nightmares? Describe them.

- Do you feel as though you are living your dharma, or are you avoiding it? Why?

- Do you have a clear idea of what your dharma is? If not, are you willing to devote yourself to finding and living it?

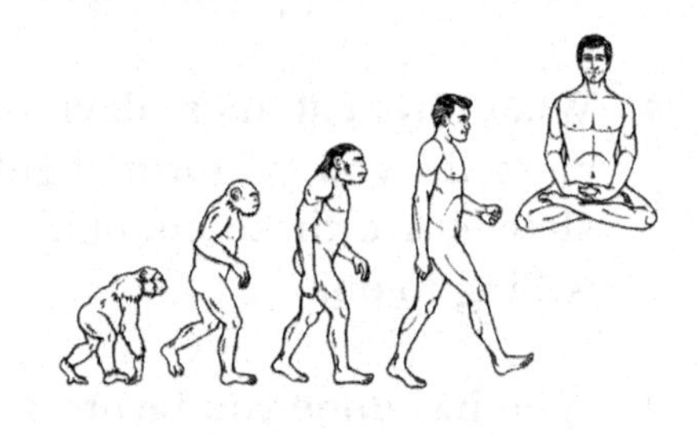

Chapter 10

Clearing the Decks

I did not wish to take a cabin passage, but rather to go before the mast and on the deck of the world, for there I could best see the moonlight amid the mountains. I do not wish to go below now.
Henry David Thoreau

You have identified your negative beliefs, set your goals, and understood your dharma. In other words, you have clearly identified the problem and what solving that problem will do for you. You have the challenge and the incentive to overcome that challenge. So: Are you ready to finally shed yourself of your old self-destructive beliefs? Changing these old beliefs allows you to embrace new ways of thinking that will interrupt your old thinking. To support these new beliefs, you're going to associate positive emotions that will break the old emotional addictions we talked about earlier. This is the single most important part of the Enlightened Savage program. I therefore want to make very sure that what we're about to do gives you the greatest possible result.

A process gets an immediate result, and a strategy maintains that result. The Enlightened Savage strategy uses several processes to change your programming. From there, I'll show you some great strategies for maintaining and nurturing your new beliefs. Whatever you do, don't shortchange the long-term strategies! Just like a baby, your new beliefs are going to be very fragile and therefore very easy to defeat. This is what happens when the "seminar high" fades. You don't want that to happen. You want the seminar high to not only stick, but grow ever stronger. That's what drives lasting growth.

Four Human Quadrants

The process of realization touches all four quadrants of human existence: emotional, logical, spiritual, and physical. Emotions are the primary stimulus that cause our thoughts, our core beliefs, to form. Thoughts tap into your spiritual sense of your mission and your place on Earth, which causes the emotions that your body is physically addicted to. These three components create your physical personal reality using the process of realization that you're already familiar with. All four quadrants must be balanced. To ignore one is to make your transformation incomplete. The best way to form the new programming is the same way you formed the old program. We'll begin by working on your emotional programming, then go on to your core beliefs. After we change your programming, I'll show you how to nurture and build your new programming as you get started on your life goals.

All that is harmony for thee, O Universe, is in harmony with me as well. Nothing that comes at the right time for thee is too early or too late for me. Everything is fruit to me that thy seasons bring, O Nature. All things come of thee, have their being in thee, and return to thee.
Marcus Aurelius

Leaving the Burrow

Be very aware that we're about to challenge your comfort zone. According to your prey brain, this is the literal equivalent of pulling you out of your burrow at high noon. Your brain doesn't want you to get killed and eaten. How can we work around this, and what will the end result be?

Stay Open

Admit that you don't know what we're going to present, even if you think you do. Keep your mind and heart open. Remember, the words, "I already know that!" are the four most destructive words ever because they can block your progress.

Focus on Opportunity

Thanks for the step up! Remember that the single biggest difference between people who achieve their goals and those who don't is that those who do forge ahead no matter what. They see their obstacles as ladders to ever-greater opportunities.

Have Faith

I need you to have faith, faith in your ability to do this and faith that we will never lead you into harm's way. Most of all, have faith that nothing will kill and eat you, even if it doesn't seem that way sometimes.

Let Go

If you would keep your soul from spotted sight or sound, live like the velvet mole; go burrow underground.
Elinor Wylie

You are carrying your old beliefs because they hold an addictive emotional charge for you. Changing your beliefs is critical, and so is letting go of the emotions that lead to and surround your beliefs. As I said before, nothing in life is ever intrinsically good or bad. Think about the worst thing that ever happened to you. The only reason it's bad is because you decided to call it bad. Illness, money issues, relationship problems, death in the family, none of these things are bad in and of themselves. Now think about the best thing that ever happened to you. The only reason it's good is because you decided to call it good. I'll say it again: Nothing that can possibly happen is ever good or bad. There is only a label. My story of the microscope is the story of how a label transformed an event. Had I applied a different label, my belief would have been much different, and my life would have turned out much differently as a result.

Be Conscious

You make choices every moment of every day by deciding how to label the things that happen in your life. The thing is, most people make those choices without thinking about it and often without even realizing it. These unconscious choices serve to reinforce your existing negative beliefs and the cycle just keeps on going and going. Learn to be conscious of everything that happens to you. Make deliberate choices about every little thing that happens. Develop the art of the "conscious interrupt" that will stop your label-making process cold and let you actively choose how to interpret the many events in your life.

Be Not Attached

A friend of mine described losing her husband to suicide and how she had spent the ensuing year "finding a new normal." I went through all kinds of emotional trauma when my marriage ended and when my financial troubles came home to roost. The same thing has doubtless occurred in your own life as well. This happens because we attach ourselves to the things in our lives the way a drowning person clings to a bit of wreckage. Afraid of what we'll see when we turn inward, we instead identify with anything and everything outside ourselves. We identify ourselves as spouses, by our professions, our politics, our likes, our physical traits, etc. and say, "I am ____," only to lose parts of ourselves when these seemingly secure things vanish. It's like clinging to a piece of wreckage that is slowly sinking.

Dust as we are, the immortal spirit grows like harmony in music; there is a dark inscrutable workmanship that reconciles discordant elements, makes them cling together in one society.
William Wordsworth

The truth is that everything we can possibly identify ourselves with is indeed sinking because nothing is permanent. Everything changes, everyone dies. Attachment, to our beliefs and to things that surround us, is the root of all suffering. Attachment leads to labeling, which leads to identification, which leads to a false sense of security. If you're paying attention, you'll know that I'm talking about the same process of realization that I've been talking about all along. I am just using some different terms that are more metaphysical and spiritual in nature.

Metaphor

Understand that this is all a metaphor. I'm using the idea of leaving a burrow at noon to describe the negative beliefs that are preventing you from coming into the full light of success and fulfillment. But do you remember when I said that we can't change our brains' wiring? We can't work against your prey instincts, so we'll work with them. We're going to change your programming to tell you that the eagle at noon is no longer the threat. Your new predator will be, for example, a cat at night.

Once we do that, you'll be out and about during the day and safely underground at night. In other words, where your programming prevented you from achieving the kind of success you need and deserve, the new programming will prevent you from not getting it. To use another metaphor, if your comfort zone represents a box surrounded by high walls, we're simply going to move the box.

What's Really Changing

They always say time changes things, but you actually have to change them yourself.
Andy Warhol

What exactly are you going to change? When this book is done, you'll still be a prey animal. Your brain will still do all in its power to avoid getting killed and eaten. The only thing you're changing is the definition of what's dangerous and not. If you replace the programming that says you need to not realize your ideal life with programming that says you deserve to do so, you will get that success. Why? Remember that you followed your negative core programming perfectly, and you can therefore follow any programming.

Focus

You'll need a quiet place free of all distractions. You want a place where you can lie or sit in total comfort.

Prepare

You'll need a pen and the questions listing your current beliefs and scores from Chapter 6. We're going to work on all beliefs that have a score or 6, 7, or higher. We'll do this by going through each belief one by one.

No Shortcuts

A cheap price is a short cut to being cheated.
Chinese proverb

Be very sure to follow each process entirely for each belief with no shortcuts. This may take a while, but compared to the number of years you've been carrying this with you, it's only an eye blink. It may get a little repetitive—and that's the whole point. You've had a lifetime of repetitive programming, so we're going to fight fire with fire. The more you hear something, the more you accept it into your personal reality. You've heard the idea that people need to see an ad nine times for it to stick? Same idea here, only this is far more important.

If you want a partner with you for support, that's totally OK. Just remember that she or he must remain quiet, with no judgment or analysis. They are there to listen and empathize.

The processes we'll use in Chapters 11-14 were created and copyrighted by Dr. Morty Lefkoe of the Decision Maker Institute. I've adapted them to fit this program, but the core processes are still the same.

A Matter of Definition

Any transition serious enough to alter your definition of self will require not just small adjustments in your way of living and thinking but a full-on metamorphosis.
Martha Beck

You've read what I have to say about negative emotions and negative core beliefs. I also said that there is no such thing as a good or bad event in our lives. Please note the distinction. The events that happen to you are never good or bad. You label them as such either consciously or by default, in accordance with your core beliefs. Once labeled, these events cause emotions that reinforce your core beliefs and perpetuate the process of realization. I call these emotions and the underlying beliefs positive or negative because they can have effects on your body that are either healthy (promote life) or unhealthy (promote death). Negative emotions can literally poison you to death. The beliefs that support your emotions must therefore be labeled positive or negative in order to match their corresponding emotions.

Again, don't be confused. All events are neutral by themselves. The emotions and beliefs that color how we perceive those events are what I call positive or negative based on their potential health effects on the human body.

Questions

Write down your answers to the following questions.

- **Are you ready to pay attention to and nurture your emotional quadrant? If not, why not?**
- **Are you ready to pay attention to and nurture your logical quadrant? If not, why not?**
- **Are you ready to pay attention to and nurture your spiritual quadrant? If not, why not?**
- **Are you ready to pay attention to and nurture your physical quadrant? If not, why not?**
- **Are you committed to leaving your old comfort zone behind and creating a new comfort zone that will empower you to live your dharma? Why or why not?**
- **Will you focus on the opportunities and keep moving forward despite the obstacles?**
- **Do you have faith that you can create the life you want and deserve?**
- **Think about some of the "bad" things that have occurred in your life. How can you see them as blessings? How can you see the "worst" thing that's ever happened to you as the biggest blessing?**
- **Think about some of the "good" things that have occurred in your life. How can you see them as obstacles? How can you see the "best" thing that's ever happened to you as the biggest obstacle?**
- **When you think about yourself, who or what do you think you are?**
- **How do you identify yourself? Answer "I am ________" as many times as needed to complete the list.**

- **What would happen to you and to your sense of self if any one or more of your "I am ________" statements suddenly ceased to be true?**

- **Are there any obstacles preventing you from moving forward? If so, what are they and how can you avoid/eliminate them?**

- **Have you made all of the preparations you need to begin changing your old beliefs?**

- **How do you feel about the processes you are about to experience? Describe your thoughts and emotions in as much detail as possible.**

Chapter 11

Changing the Programming

This is it! This is where your life begins to change forever, where you become the person who will live the goals you listed in your workbook, the person who will fulfill your dharma. Doing this will open you to experiencing joy, fulfillment, love, and wealth (material or otherwise) beyond measure. So let's get started! Remember that these processes are not intended as therapy. If you find yourself approaching any sort of crisis state, stop and seek appropriate assistance.

Setting the Stage

The stage is not merely the meeting place of all the arts, but is also the return of art to life.
Oscar Wilde

If you're going through this chapter for the first time (or if you've paused in between changing beliefs), have a seat and get comfortable. If you're repeating this chapter immediately or at any time after going through a previous round, please remain sitting comfortably, and make sure that you follow this exercise thoroughly each time you go through this chapter. Remember: There are no shortcuts!

We'll begin with an easy relaxation exercise. Taking the time to relax will help you change your negative mental programming Relax and take gentle even breaths for a few seconds. Hold your head steady and focus on any spot in the room as you clear your mind.

Take a deep breath. Hold it as long as you can. As you slowly exhale, feel a wave of relaxation starting at the top of your head all the way down to your toes. Relax. Another deep breath, hold, relax. Do it a third time. Imagine yourself living your ideal life, fulfilling your dharma, and reaping all of the rewards you deserve.

What do you see? What are the colors, the textures, the shapes? What do you hear? What do you smell? Flowers? Saltwater? Pine needles? Something else? What can you taste, and what can you feel as you reach out and touch your surroundings? Immerse yourself completely in this wonderful, fulfilling world.

You are taking an enormous step that takes great courage, strength, and faith. That means that you are brave, strong, and faithful. Clasp your hands together as if you were shaking them and congratulate yourself out loud in your own words. Great job. Good for you!

You deserve to make this change. You are a unique person in this vast universe. That makes you infinitely valuable and infinitely deserving of success. Tell yourself out loud in your own words how much you deserve this change, and how wonderful it is to be freeing yourself from this belief.

Life is an adventure in forgiveness.
Norman Cousins

How do you feel in this moment? What good emotions are you aware of? Do you feel happy, excited, purposeful? Any good feeling you have is okay. Be sure to voice your feelings aloud.

Take this wonderful feeling and bring your thumb and your forefinger together; hold that feeling right there. Whenever you're feeling frustrated, or sad, or depressed, or angry, or hopeless, or confused, you can bring these fingers back together and immediately reconnect with the positive emotion that you feel right now.

Now let yourself think about some of the things that you choose to see as accomplishments you've made in your life. It could be anything. Maybe it's a science fair award when you were in school. It could be a promotion, a job, a business, anything. If you're repeating this chapter, think of something different each time. Clasp your hands together. Hear yourself say out loud "Good job! Way to go! I did this! I feel good about this, and I deserve to feel good about this! I can do anything I want!" These accomplishments are not good by themselves, nor do they define you in any way. They are good because you did them and because you saw and labeled them as good. Your body may be what you eat but you are not what you do.

Forgiveness and Thanks

To give thanks in solitude is enough. Thanksgiving has wings and goes where it must go. Your prayer knows much more about it than you do.
Victor Hugo

OK. Pick up your answers to the questions from Chapter 6 and select the first (or next) belief rated 6, 7, or higher to change. You should already identified this belief, where it's coming from, and how it's hindering the success you want and deserve. Imagine how wonderful you'll feel when this huge weight is taken off your shoulders.

Take a moment to reconnect with this belief and any associated emotions. These emotions could come from when you formed the belief or from its effects. You may feel sad, angry, or frustrated. That's OK.

Forgiveness is terribly important. You must forgive everyone in your life who instilled or contributed to this belief. They did it because of their own programming, and never intended to hurt you. There's an ancient story about rich merchants dumping bags of gold in the Temple offering box. Then a poor woman gives two cents. The merchants only gave a little of their money, but the woman gave all she had. No matter how much or how little, everyone in your life gave all they had. Forgive them. Say whatever words of forgiveness you like out loud. If you've already forgiven these people, do it again. There's no such thing as too much forgiveness.

Thank the belief for existing. Thank yourself for following it so faithfully. This belief made you who are today by helping you become a unique person who deserves every success in the world. This belief helped bring you here to this place now. It may not be the best belief and that's OK because it still served you as well as it could. Say whatever words of thanks you like to this belief out loud as if it were a person. Thank your brain for following this belief so well that you now know you can follow any programming at all.

From Effect to Cause

The most violent revolutions in an individual's beliefs leave most of his old order standing. Time and space, cause and effect, nature and history, and one's own biography remain untouched.
William James

Identify some negative pattern of behavior, feeling, or circumstance that exists in your life because of this specific belief. If what you come up with involves more than one belief, that's fine; we'll address the rest later.

It's possible that this pattern might not fit any listed belief. That's OK. If this happens, take a moment to find a previously unlisted belief and score it like the other ones. If that belief is strong enough, let's change it when its turn comes.

This belief might not fit a definite pattern, but you might still have a gut sense of believing it anyway. If that's true, set this particular belief aside. We'll cover it in the following chapters. For now, return to the beginning of this chapter and start over with a different belief. There is absolutely nothing wrong if this happens. All we're doing is using the correct process for different types of beliefs. You've just taken a step towards doing that right. Keep going. You're doing great!

Let's look at where this belief came from. Read your notes about what you saw, heard, or experienced that created or reinforced this belief. How does this feel? Have you nailed down where this belief came from?

If not, allow yourself to remain in your relaxed state and see if you can relive the times when this belief was formed. Make any changes as they come to you. Remember that you don't have to feel bad. You deserve to replace this belief and deserve to make sure that you're making a clean sweep. All we're doing is double-checking.

Great job. By finding all the roots, you can fix all the fruits. Remember that your brain might be putting up huge obstacles in your path. If so, thank it out loud for protecting you and for keeping you alive. It is, and always has been, doing its job perfectly!

Alternate Realities

If all the world's a stage, I want to operate the trap door.
Paul Beatty

Now that you've identified the source of the belief, take a moment to congratulate yourself for having the courage to face this belief and dig into it. Try to separate whatever you saw, heard, and experienced from the belief itself. Pretend you're seeing the same things for the very fist time right now. Are you coming up with anything different?

Made-up Truth

Error has made animals into men; is truth in a position to make men into animals again?
Friedrich Nietzsche

This belief that's been running your life for this long and that you accept as the truth, is nothing more than a made-up truth. Something happened, and you formed a conclusion that has been ruling your life ever since. As you look back on this belief, can you come up with any alternative explanations? If you interpreted something as negative, could you now see it as truly positive? In other words, do you understand that this belief which you hold as the truth is nothing more than one of many possible truths? What other interpretations can you form from the same events that formed this belief?

Since the physical world you experience as your personal reality is nothing but an effect, isn't the meaning that you think is inherent in what you saw, heard, and experienced only real inside your mind, and not in the events themselves? Can you accept the fact that your belief is nothing more than the result of the interpretation that you made of what you saw, heard, and experienced? If not, that's OK; return to the beginning of this chapter and try it again.

Did what you saw, heard, and experienced support the interpretation that led to this belief? In other words, did you create a reasonable response to the event, or events, that formed this belief? How old were you when this belief formed? Would most people that age probably reach the same conclusion you did? What about people your current age?

Many Possibilities

It is a world of startling possibilities.
Charles Fletcher Dole

Keep going! Come up with as many different interpretations as you can think up until you fully accept the idea that your conclusion, your current belief, is nothing more than one of many possible beliefs that you could have formed—and that each of those possible beliefs is equally as valid as the one you did form. Your belief is one logical interpretation for what you saw, heard, and experienced. Now that you've created alternate interpretations of the same events, is the belief you formed at the time the truth, or only one interpretation of what you saw?

When you formed this belief, did it seem real, as if you'd discovered something really true? Probably, or you wouldn't have this belief. Question: Did you ever really see this belief itself in the world? Probably not. So all you ever saw were the sights, sounds, and experiences that formed the belief.

By themselves, what do these events signify? Nothing! What inherent meaning do they convey? None! Out loud, say the belief you've been carrying and then say the following: "This belief carries no meaning but the meaning I give it. I create my own personal reality, not this belief. The event(s) that caused this belief are neither positive nor negative. Whatever good or bad I see in them only exists because I chose to see it. I have the power to label anything in my life however I choose."

If this belief was never out there in the world, where has it been all these years? In your head, inside your thoughts, feelings, and actions that created your results. Do you now see and accept as reality the idea that the only place your belief has ever truly existed is as one interpretation out of many possible interpretations inside your mind? How does this realization feel to you? What emotions come to you as you ponder this idea? Say them out loud and make them real. Say "I accept that my belief is only one interpretation of many possible interpretations."

Forming a New Belief

In this moment, do you still carry this old belief? If so, that's OK. Go back to the beginning of this chapter and start again. This book is here for you as many times as you need it. I hope you are totally OK with the idea that some powerful beliefs need to be chipped apart. Your beliefs have taken a lifetime to sink in, and some will not go as willingly as others.

You must put fear out of your mind. Confront it with the belief that the past is over with, that a new life lies ahead.
Edward T. Lowe

If this old belief is gone, what's replaced it? Do you feel an exciting sense of possibility? Do you think that you might actually be able to do things that were inconceivable just a few minutes ago? In other words, is your old belief true for you any more? State your old belief out loud: How does saying those words feel? If you said false, unreal, or anything similar, then congratulations; the belief is gone! Take a few deep breaths and feel the sense of freedom and power washing through you. Repeat several times: I am free. I am free!

Go ahead and pick a positive interpretation of the events that caused your negative belief. Do you see that this interpretation is just as valid as your old one? Based on this fresh look, what is your new belief? Say it out loud, beginning with "I believe" followed by your new belief. How does it feel to have this new empowering belief? Voice those feelings aloud. Laugh, cry, shout if you want. Celebrate! Congratulate yourself. You did it. You did it, and if you can change your programming, you can do anything, within the laws of physics. Most of all, you must understand that action is the key to making new beliefs real. No amount of "attraction" or "manifestation" will do that for you, no matter how hard you try. That is a mathematical fact.

Success is a process, a quality of mind and way of being, an outgoing affirmation of life.
Alex Noble

Affirming the New Belief

Question: Has your life been totally consistent with your old belief? Based on what you learned about the process of realization, can you also see that choosing any of the other interpretations you just named would have made your life consistent with those interpretations? Based on this, do you see and accept the reality that your life will be totally consistent with your new belief? If so, promise yourself right now that you will be true to your new belief. If you have a partner with you, look her or him in the eye when you make this pledge. If not, that's OK; go back to the beginning of this chapter and begin again.

Let's take a look at your life in the context of your new belief. Find events that are consistent with your new belief. Write your new belief and anything you've seen, heard, or experienced that supports this new belief.

Congratulations! You have just shed a huge burden. It's OK to feel giddy and excited. Dump a heavy suitcase and you'll feel like you can leap a tall building—and guess what: You can! You are the decision maker, the creator of your beliefs, your inner sunglasses that reveal themselves as your behavior and everything that you see, touch, hear, smell, and taste. Is it real to you that you create your beliefs? Is it real to you that your beliefs determine your life? If you create the beliefs that create your life, what does that make you? The creator of your own life. You create the creation. You create yourself every minute of every day.

You have just eliminated a belief that has been slowing you down for years. Eliminating every one of your remaining negative beliefs is just as easy or even easier.

Onward and upward.
Abraham Lincoln

On to the Next Belief

It is very important that you follow this process for every negative belief. Yes, you may be working in this chapter for a long while, but how long have those old beliefs ruled you? This small investment will begin paying off. Not next year not tomorrow, but right now—and the payoff will continue for the rest of your life!

Whenever you're ready, go back to the beginning of this chapter and change the next belief that you scored a 6, 7, or higher. If you're not ready, that's OK. Go enjoy your new belief. Start absorbing it. Come back for the next one when you're ready. Just please make sure not to go on to the next chapter until you've eliminated all of your negative beliefs rated 6, 7, or higher for which you can identify specific causes.

If you get stuck, that's OK. Move on to different beliefs and circle back when you're ready. Never forget to be gentle, patient, and loving with yourself.

An Important Note

Never forget that your new beliefs are no more or less truthful than your old beliefs. The difference is that positive beliefs will create positive emotions, which will attract more positivity into your life, which will help you get where you want to go and live the kind of life you want to live, by motivating you to take the correct actions towards achieving your goals and living your dharma. Achieving the success you deserve absolutely requires action to realize your dreams. Anything less is a complete and total waste of time, period.

Questions

Congratulations! You have just taken a major step on the road to freeing yourself from your old beliefs and stepping into who you are meant to be. Be sure to take the time to savor your new feelings and revel in this huge success. Your beliefs are the root of your personal reality. Changing this root will change your entire life.

- **Have you addressed every belief you rated a 6, 7, or higher in Chapter 6? If not, go back and repeat this chapter once for every one of your remaining negative beliefs.**

- **How does it feel to have changed your old negative core beliefs and replaced them with positive ones? Describe your thoughts and emotions in as much detail as possible.**

- **Do you understand that achieving your goals requires action and not attraction?**

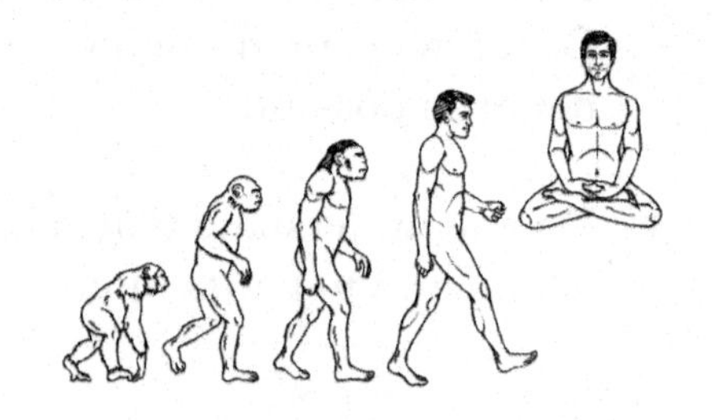

Chapter 12

Eliminating Emotional Triggers

The most beautiful emotion we can experience is the mystical. It is the power of all true art and science. He to whom this emotion is a stranger, who can no longer wonder and stand rapt in awe, is as good as dead.
Albert Einstein

Negative emotional triggers cause you to experience negative emotions that sap your energy and either keep you from taking action (at best) or even compel you to take the wrong action (at worst). You've just freed yourself from negative core beliefs and already know how wonderful it feels to liberate yourself. Eliminating emotional triggers is just as easy, so let's keep going.

This chapter is only for negative emotions that are triggered by certain events or other stimuli. These may or may not involve beliefs. If a belief is involved and you have been unable to identify it, then it may be buried or otherwise repressed. If not, then these triggers function like *operant* (behavioral) conditioning where you have developed a reflex response to one or more stimuli. A famous example of such conditioning is Ivan Pavlov's dog, which learned to salivate in anticipation of a meal when a bell was rung. This chapter helps you with your conditioning. If you still have any negative beliefs with identifiable causes, please go back to Chapter 11.

The same rules apply as before: Go through this chapter once for each of any negative emotions you may experience as a result of specific stimuli. If you get stuck, move on to the next trigger and circle back later. Never forget that these processes are intended to help you unleash your inner greatness, not to treat any disease or disorder. They are not intended as therapy.

Let's begin by laying some groundwork: What negative or unpleasant emotion do you want to eliminate? What specific trigger causes this emotion? When did you first experience this emotion with this stimulus? If you have no specific triggers, then this emotion may be caused by a belief and should be eliminated by either going back to Chapter 11, or by going ahead to Chapter 13 when the time is right (but not before).

You've just either gotten through Chapter 11 or are repeating this chapter. At this point, you're probably on an emotional high... and rightly so! You have accomplished more than you might be aware of right now. Be aware, however, that this high may not last. Even after freeing yourself from your negative beliefs, you might continue experiencing negative emotions caused by triggering events that always seem to know every one of your buttons. I, for one, believe that most, if not all, negative emotions trace their roots back to the primal fear of being killed and eaten. These triggering events are somehow tapping into that fear buried deep inside all of us. You may not know exactly where it's coming from, and that's OK.

Setting the Stage

Pick one negative emotion and the events that trigger it. Do you think this emotion stems from the perception that your survival is threatened? Now ask yourself this: Are you being threatened right here, right now? Probably not. You see, the perceived threat is rarely present when the fear is triggered. The perception of threat comes from interpreting early life events as dangerous. This is your prey animal brain at work.

The meeting of preparation with opportunity generates the offspring we call luck.
Anthony Robbins

These early events usually involve your parents and other members of your immediate group, the very people upon whom your survival depends. If you ever experienced your parents' love and/or group acceptance as anything but unconditional, the instant question—and underlying fear—is how the heck you're going to survive if you can't count on them. It's a perfectly valid question.

Our fears whisper to us in many different ways by causing doubts, frustration, sadness, defeat, excuses for not taking action—the list goes on and on and on. Your emotions keep coming up in various forms whenever you experience a stimulus that challenges the deeply buried underlying belief or beliefs.

If you're reading this chapter for the first time (or if you've paused in between eliminating triggers), have a seat and get comfortable. If you're repeating this chapter immediately after reading it last time, please remain sitting very comfortably. Remember to go through this process completely every time with no shortcuts.

Relax and take gentle, even breaths for a few seconds. Hold your head steady and focus on any spot in the room as you let all of your cares and worries slip away. Take a deep breath. Hold it as long as you can. As you slowly exhale, feel a wave of relaxation starting at the top of your head all the way down to your toes. Relax. Another deep breath, hold, relax. Do it a third time. Imagine yourself living your ideal life.

As we free our breath (through diaphragmatic breathing) we relax our emotions and let go our body tensions.
Gay Hendricks

What do you see? What are the colors, the textures, the shapes? What do you hear? What do you smell? What can you taste, and what can you feel as you reach out and immerse yourself completely in your surroundings? Let yourself experience all five senses. Remember, parts of your brain can't tell the difference between processing dreams and waking, making both equally real.

You are taking another enormous step that requires more courage, strength, and faith. You've just eliminated your negative beliefs. You've proven that you are indeed brave, strong, and faithful. Take a moment to clasp your hands together as if you were shaking them and congratulate yourself out loud. You're awesome!

You've come this far. You deserve to go the distance, to become an enlightened savage. You are infinitely unique, valuable, and worthy. Tell yourself out loud how great you are for having come this far, and how liberating it will be to lose these negative emotional triggers.

What good emotions are you aware of right now? Happiness, fulfillment, joy, purpose? Any good feeling you have is okay. Again, voice your feelings aloud and hear yourself saying the words. Immerse yourself in these wonderful feelings.

Think of an accomplishment, no matter how minor, that you haven't thought of yet. Clasp your hands together. Hear yourself say out loud "I did this! I deserve to feel good about this, and I do feel good. In fact, I feel wonderful! I can do anything I want to do!"

Think of specific events in your life that bring on negative feelings as if they were pulling a trigger. Imagine how wonderful you'll feel when you free yourself from this baggage. Take a moment to reconnect with these triggers, and with the emotions they stir up time and time and time again. Let yourself feel whatever you feel when these triggers occur.

Forgiveness and Thanks

As with your beliefs, forgiveness is absolutely critical. If any person or persons instilled or contributed to your emotional triggers, forgive them. Forgive yourself for responding the way you did. No one set out to hurt anyone else. We are all prey animals doing our best to make our way in this Universe. No matter how much malice or evil you may have sensed or experienced, chances are that it only came from fear.

One of life's gifts is that each of us, no matter how tired and downtrodden, finds reasons for thankfulness: for the crops carried in from the fields and the grapes from the vineyard.
J Robert Moskin

The worse you were treated, or the worse you treated yourself, the worse the fear. Voice any words of forgiveness you like out loud. If you've already forgiven these people, do it again. Can you feel love for them and for yourself? Every trigger and every feeling helped get you to this point here and now. Thank them.

Thank these triggers for existing and thank yourself for following them so faithfully and for feeling the way you did. These emotions make you a unique person who deserves every success in the world. They helped bring you here to this place of change. They have served you as well as they could. Thanks your emotions out loud as if they were people. And thank your brain for feeling these emotions so well that you know you can feel any emotion you want to.

From Effect to Cause

Shallow people believe in luck and in circumstances; strong people believe in cause and effect.
Ralph Waldo Emerson

You need to understand that the stimuli that trigger your emotions aren't the root cause behind those emotions. The emotions themselves come from the meaning you gave to the original cause; the current stimuli just happen to be associated with that cause. The the original cause is only negative because you decided it was; it was neither good nor bad on its own.

As we proceed, you're going to notice some strong parallels between removing your stimuli and removing your negative core beliefs. This is no accident! As before, we'll repeat these processes once for each specific or generalized emotion you wish to eliminate. You may do this alone or with a partner. If you have a partner, be sure to follow the rules of silent non-judgment and confidentiality. As before, if you get stuck, move on to another trigger and come back to this one later.

This emotion may not be connected to your core beliefs. Remember that original emotional programming leads to core beliefs, but that may not account for everything.

Question: Specifically who, what, where, when, why, and how did these earlier experiences cause the emotion associated with the stimulus? Can you see that the only reason the stimulus causes the emotion today is that you never distinguished between the real cause of the emotion and the current stimulus? In other words, can you see that your emotion was never caused by the stimulus but only by the original cause? Also, do you understand that the original cause was not good or bad by itself, that you chose (consciously or not) to label it as such?

Alternate Realities

There are some people who live in a dream world, and there are some who face reality; and then there are those who turn one into the other.
Douglas Everett

To make this distinction real, if the circumstances earlier in your life that originally caused the emotion had been different—if the opposite of the original cause had happened instead—would the current stimulus still cause the same emotion? If the original cause had happened but you had chosen to see it differently, would it still cause the same emotion? If the stimulus did not cause the emotion then, would it cause the emotion now? No. Can you therefore see that the stimulus is not real? Close your eyes and imagine the current stimulus.

Let it permeate your body. Do you still feel the emotion? If so, that's OK; go back to the beginning of this chapter and start over. It may take a few passes to completely defuse an emotional trigger that has been with you for many years.

Breaking Free

The only risk in bondage is breaking free.
Gita Bellin

Congratulations! You've just shed another huge burden. If you have a partner with you, share your feelings with her or him. As before, It's OK to feel giddy, and excited, and overcome with emotion.

Dropping these heavy weights will let your spirit fly, and will make the success you seek possible. You create your beliefs that, in turn, create everything that you see, touch, hear, smell, and taste. This makes you the creator of your own life, the creator of the creation! You have just eliminated a negative emotional trigger that has been slowing you down for years. Eliminating every one of your remaining triggers is just as easy! All that is required is for you to take the correct action to realize your liberation.

Now, if you're ready, go back to the beginning of this chapter, and let's change the next emotional trigger. If you're not ready, that's OK. Start absorbing your freedom from the trigger you just eliminated, then come back for the next one when you're ready. Just make sure not to go on to the next chapter until you've eliminated all of your negative emotional triggers.

Questions

Congratulations! Completing this chapter marks another major step on your road to becoming an enlightened savage. Allow yourself to take some time to savor your success. Your emotions are the very root of the process of realization. Change the root and change your entire life.

- **Have you addressed every negative emotional trigger? If not, go back and repeat this chapter once for every one of your remaining negative emotional triggers.**

- **How does it feel to have defused the triggers that spawned negative emotional reactions? Describe your thoughts and emotions in as much detail as possible.**

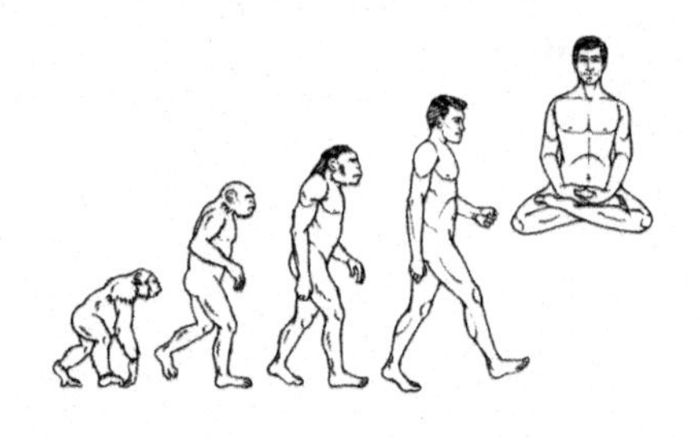

Chapter 13

Quieting Nagging Doubts

A person who doubts himself is like a man who would enlist in the ranks of his enemies and bear arms against himself. He makes his failure certain by himself being the first person to be convinced of it.
Alexandre Dumas

This chapter is designed for those nagging negative senses, that deep disquiet that sometimes inhabits our bodies without any specific trigger. Again, the same rules apply: Use this process once for each of any negative senses you may have hanging over you. A partner is perfectly OK so long as you remember the rules. You've just freed yourself from negative core beliefs and removed emotional triggers. Quieting your nagging doubts is just as easy. You are well on your way to true emotional freedom and to building a whole new personal reality for yourself. I am very proud of you, and I hope you're just as proud of yourself, because you deserve it!

Even after eliminating negative core beliefs and defusing emotional triggers, you may still harbor some deep discomfort. Take a good look inside and find your sense of self. Don't worry about finding the right description or explanation. Just experience that sense as fully as you can right now. Remember: This process is not intended to provide any sort of therapy.

Setting the Stage

> *The world is a stage and most of us are desperately unrehearsed.*
> Sean O'Casey

Make sure you're sitting comfortably. Once again, relax and take gentle, even breaths for a few seconds. Hold your head steady and focus on any spot in the room as you let all of your cares and worries slip away.

Take a deep breath. Hold it as long as you can. As you slowly exhale, feel a wave of relaxation, starting at the top of your head all the way down to your toes. Relax. Another deep breath, hold, relax. Do it a third time. Imagine yourself right back in your ideal, extremely successful life. What do you see, hear, feel, smell, and taste? Feel all of your senses tuning into this reality.

Look at you: You have eliminated negative beliefs and quieted emotional triggers. You have achieved something that most people never even attempt. Congratulate yourself out loud in your own words. Clasp your hands together as if you were shaking them. You deserve this. You deserve this! Tell yourself out loud how great you are for having come this far, and how close you now are to becoming an enlightened savage.

What positive emotions are you feeling right now? Voice those feelings out loud. Hear yourself saying the words and let yourself really feel these great feelings.

Think of any accomplishment that you haven't thought of yet. Clasp your hands together. Hear yourself say out loud "I did this! I deserve to feel good about this, and I do feel good. In fact, I feel wonderful! I can do anything I want to do!"

Forgiveness and Thanks

> *Heraclitus says that Pittacus, when he had got Alcæus into his power, released him, saying, "Forgiveness is better than revenge.".*
> Diogenes Laërtius

These nagging senses you've carried with you have served you well. They helped get you to this point today. Forgive yourself for having them. You have always done the very best that you know how to do. Voice any words of forgiveness you like out loud. Thank your nagging senses for bringing you to this point right here and now.

Thank your functioning brain for following your doubts and fears. It followed your programming perfectly. That means you can follow any programming just as perfectly.

From Effect to Cause

To design the future effectively, you must first let go of your past.
Charles J. Givens

Here again, there are many parallels between the exercise we're about to perform and the questions in previous chapters. Humans are amazingly consistent creatures. What works in one area can be modified to serve in many areas. Using similar, consistent methods to remove beliefs, emotional triggers, and nagging senses instills the processes into your long-term memory, where they will help you keep that old programming from returning. Any time you find yourself sliding back, you'll remember the process and snap right back out of it. You may not remember the exact steps. That's OK; you will remember the basics, and those are more than enough.

Think back to any event that might have caused this fear, doubt, or other nagging sense to form within you. Immerse yourself as much as possible in that moment and relive it. Make it fully real with all of your senses.

Now that it is real, put it into words. What is the earliest event you can remember that caused this sense?

Alternate Realities

Reality leaves a lot to the imagination.
John Lennon

Can you see that the only reason that you feel this sense as inherent in you today is that you never distinguished between yourself and the specific external circumstances that really caused the sense?

In other words, can you see that the sense was never inherent in you? Do you understand that it came from something outside you? To make this distinction real, if the circumstances that originally caused the sense had been different, if the opposite of what happened had actually occurred, would you have had the sense again? Had you chosen to interpret what happened differently, would you have ever had this sense? If you didn't have this sense then, would you have it now? Look inside. Do you still experience yourself as having this sense? No. Where did it go? More to the point, where has it been? Nowhere!

Breaking Free

Freedom is knowing who you really are.
Linda Thomson

Congratulations! You've just freed yourself from yet another burden. If you have a partner with you, share your feelings with her or him. As before, It's OK to feel giddy, excited, and overcome with emotion.

Dropping these nagging senses will let you move and live with confidence and eagerness. You alone create your own life! You have just eliminated a nagging sense that has been slowing you down for years. Eliminating every one of your remaining negative senses is just as easy. As always, action is the key.

Now, if you're ready, go back to the beginning of this chapter, and let's change the next negative sense. If you're not ready, that's OK. Start absorbing your freedom from the sense you just eliminated. Come back for the next one when you're ready. Just make sure not to go on to the next chapter until you've eliminated all of your negative senses.

Questions

Congratulations again! This chapter marks still another major step on your road to becoming an enlightened savage. Allow yourself plenty of time to relish your success. Freeing yourself of nagging doubts allows you to forge ahead with confidence.

- **Have you addressed every nagging doubt and insecurity? If not, go back and repeat this chapter once for every one of your remaining negative emotional triggers.**

- **How does it feel to have laid your nagging doubts to rest? Describe your thoughts and emotions in as much detail as possible.**

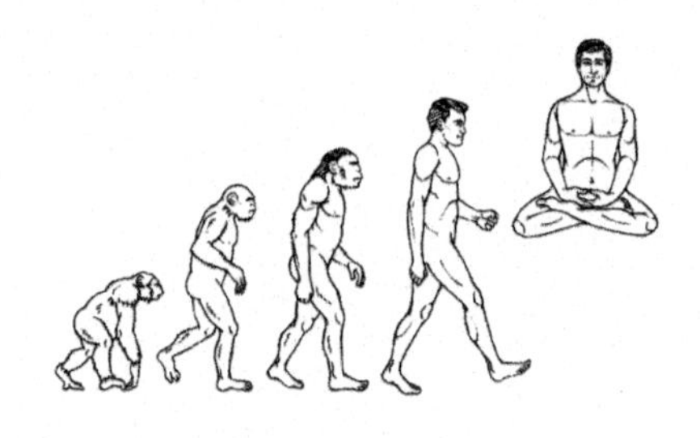

Chapter 14

Expect the Best

Winners make a habit of manufacturing their own positive expectations in advance of the event.
Brian Tracy

You've eliminated your negative core beliefs, emotional triggers, and non-specific doubts and fears. The next step is to make sure that your new expectations are in line with your new programming. This is the last piece of the puzzle, because your behavior and emotional state all too often depend on your expectations. Think of expectations as forward-looking beliefs that govern how you will experience life. Remember the process of realization? Let's make sure you expect—and therefore build—the best possible personal reality.

What is a negative expectation and how does it work? Here's an example we can all identify with: Too many of us expect life to be difficult, We expect to not get what we want and deserve. Toss in a few negative beliefs, and we find ourselves resisting setting goals and giving up as soon as we experience any difficulty. We end up drifting through life with no real purpose. We've talked about your goals and your dharma. Now it's time to make sure that you expect your goals to come true and to fulfill your dharma.

Setting the Stage

First comes thought; then organization of that thought into ideas and plans; then transformation of those plans into reality. The beginning, as you will observe, is in your imagination.
Napoleon Hill

Begin by writing down your negative expectations. Use your stated goals as a starting point. Do you expect to accomplish them? Do you expect this to require lots of struggle?

If you're reading this chapter for the first time, or if you've paused in between reruns, have a seat and get comfortable. If you're repeating this chapter immediately after reading it last time, please remain sitting very comfortably. Relax and take gentle even breaths for a few seconds. Hold your head steady, and focus on any spot in the room as you let all of your cares and worries slip away.

Take a deep breath, and hold it as long as you can. Exhale slowly, feeling a wave of relaxation start at the top of your head and wash all the way down to your toes. Relax. Take another deep breath, hold it, and relax. One more time. Imagine yourself right back in your successful rewarding life.

What do you see, hear, feel, smell, and taste? Plunge into this fulfilling world. Clasp your hands together and congratulate yourself for coming this far. You did this. You deserve all the credit and all of the rewards!

Let yourself realize how much you deserve to expect the very best in your life. You have infinite value and worth. Tell yourself how valuable you are out loud. Say how great it will be to always expect the absolute best out of life. Tell yourself out loud how much you deserve to expect—and get—the best.

How do you feel right now? Let any good emotion surround and permeate you. Go ahead and voice your feelings aloud.

Think about more of your accomplishments. Be sure to think of something new, especially if you're repeating this chapter. Clasp your hands together and congratulate yourself out loud. The fact that you did this means you can do anything! Anytime you doubt this, you'll be able to clasp your hands together and immediately reconnect with this success.

Forgiveness and Thanks

You have many choices. You can choose forgiveness over revenge, joy over despair. You can choose action over apathy.
Stephanie Marston

These negative expectations you've carried with you have served you well. They helped get you to this point today. Forgive yourself for having them. You have always done the very best that you know how to do. Voice any words of forgiveness you like out loud. Thank your negative expectations for bringing you to this point right here and now.

Thank your functioning brain for following your doubts and fears. It followed your programming perfectly, and that means you can follow any programming just as perfectly.

From Effect to Cause

Ideas are the roots of creation.
Ernest Dimnet

Let's begin with the first negative expectation you wrote down a few minutes ago. Take a few moments to imagine what might happen—or not happen—if this expectation becomes true. Allow yourself to experience the sense of failure and defeat, sadness, frustration, and anger you will feel if this comes to pass.

Why do you have this negative expectation? What are you scared of? That you can't or won't do any better? Or is it something different perhaps? Marianne Williamson has this to say about expectations:

> *Our deepest fear is not that we are inadequate. Our deepest fear is that we are powerful beyond measure. It is our light, not our darkness, that most frightens us. We ask ourselves, Who am I to be brilliant, gorgeous, talented, fabulous?*

Does this sound like a prey animal scared of getting killed and eaten? Like a person afraid to leave their comfort zone because of negative core beliefs? Marianne goes on to ask a very simple, yet infinitely profound, question:

> *Who are you not to be brilliant, gorgeous, talented, and fabulous?*

The fool saith, I have no friends, I have no thanks for all my good deeds, and they that eat my bread speak evil of me.
Ecclesiasticus 20:16

In fact, you already are all of these things. Why? Because you're unique in all the Universe, and therefore have infinite intrinsic value. Gold doesn't have to do anything special. It's valuable just because it exists. Same with you. You exist. Therefore, you are valuable beyond measure.

You may beat yourself up because of this negative expectation. If so, then it's time to forgive and thank yourself for carrying this expectation. You've followed it to the letter, which means you can follow any expectation at all. That expectation exists to keep you from being killed and eaten. The fact that you're alive to read this book means that it worked. Forgive and thank yourself out loud for carrying this expectation with you. It has served you well. So will your new expectations.

Alternate Realities

Again, please read this chapter once for each negative expectation you identified earlier. If you have a partner, great. Just please remember that your partner must offer silent, confidential support. Ready to lose this expectation? Here we go.

> *What if all ponds were shallow? Would it not react on the minds of men? I am thankful that this pond was made deep and pure for a symbol.*
> Henry David Thoreau

What specific issue or challenge does your negative expectation concern? What happened in your life that led you to form this expectation? Can you see that your expectation made sense given the many experiences you had that were similar to what you just described? Can you see that your expectations are a function of those specific events? Can you see that the specific issue behind this expectation never created your expectation?

What if the events of your life had been very different? What if, for example, the events that led you to this expectation had been just the opposite? Would you—could you—have formed your same expectation then? Do you understand that your current expectation for this issue would be totally different if your earlier experiences had been different?

Next, describe the differences between your earlier circumstances and today's circumstances. Can you see that today is very different from the earlier time that led to your expectation?

Don't discuss what you want, what you wish for, or how you'd like it to be. Focus solely on what any reasonable person would expect given your life today. Look inside. What do you expect right now? Is this new expectation different than the old one? Do you accept that you have the power to make this new expectation real by building whatever personal reality you like?

Breaking Free

You take on the responsibility for making your dream a reality.
Les Brown

Congratulations! You've just freed yourself from one of the most insidious traps of all, your own negative expectations. If you have a partner with you, share your feelings with her or him. As before, It's OK to feel giddy, excited, and overcome with emotion.

Dropping these negative expectations will let you build a much more positive reality, because you'll be expecting the best from the outset. You alone create your own life! You have just dropped a huge anchor that has been slowing you down for years. Eliminating every one of your remaining negative expectations is just as easy, provided that—you guessed it—you take the right actions toward realizing your goals.

Now, if you're ready, go back to the beginning of this chapter. Let's change the next negative sense. If you're not ready, that's OK. Start absorbing your freedom from the sense you just eliminated, and come back for the next one when you're ready. Just make sure not to go on to the next chapter until you've eliminated all of your negative senses.

Questions

What a difference a few chapters makes! By now, you have changed your negative core beliefs, eliminated negative emotional triggers, cast aside doubts, and now have discarded your negative expectations. Make very sure to take enough time to truly relish your accomplishments. By no longer expecting bad things to happen, you're priming yourself to expect the very best out of life. And since you create your own personal reality...

Write your answers to the following questions:

- **Have you addressed every negative expectation? If not, go back and repeat this chapter once for every one of your remaining negative emotional triggers.**

- **How does it feel to be free of the negative expectations that all too often cause bad things to happen? Describe your thoughts and emotions in as much detail as possible.**

Chapter 15

Consciousness

> *Ask not what the world needs. Ask what makes you come alive... then go do it. Because what the world needs is people who have come alive.*
> Howard Thurman

So far, we've learned how our prey brain works and how our core beliefs affect our entire lives. You know that nothing that happens to you is ever intrinsically good or bad unless and until you choose how to interpret and label it. You also know that you are always labeling your experiences, either consciously or unconsciously. We then went through several processes designed to clear your negative beliefs, eliminate emotional triggers, and quiet your doubts. I've also explained that your new beliefs and mental programming are very new. Like all newborns, they are vulnerable and require constant protection and nurturing in order to grow and flourish.

The remainder of this book contains many different strategies to help you accomplish all this. First, though, we need to build the foundation for those strategies by constructing the appropriate mental framework. This framework consists of six components:

- Understanding your fundamental needs,
- Resourcefulness,
- Letting go,
- Non-attachment,
- Conscious choice, and

- Developing a "conscious interrupt" that watches over you and intervenes to give you the opportunity to make conscious decisions.

Once you understand the framework, I'll show you how to use it to support your ongoing growth in following chapters.

No amount of time can erase the memory of a good cat and no amount of masking tape can ever totally remove his fur from your couch.
Leo Buscaglia

Let's use the computer analogy to understand why this framework is so important for your ongoing growth: Yes, you've deleted a lot of old programming, but that programming must be replaced. Simply deleting the programming merely removes it from active use, but does not completely erase it. When you delete a file on your computer, the information remains on your hard drive. All the computer does is mark that space as available for possible use when you save a new file to that drive. You can buy programs to undelete files, which has saved me more often than I'd care to admit. Your brain contains a powerful undelete feature. This is great for preserving and recalling memories, but comes with potential side effects: Unless you actively overwrite the old programming, you will find it creeping back into use. To use another analogy, your old programming may be knocked out, but is a long way from being completely gone.

The six-part framework I'm going to share with you is the mental approximation of formatting a hard drive. Formatting a hard drive prepares it to hold information by specifying how the drive will hold and catalog that information. In other words, the file format tells the computer how to send and receive information to and from that hard drive. For you, the six-part framework is the format you will use for all of the strategies that follow. This is why it's so important.

Fundamental Needs

A man travels the world over in search of what he needs and returns home to find it.
George Moore

In Chapter 1, you learned about the six fundamental goals of life, which are (in no particular order): money, love, health, longevity, fun, and work. Each of your core beliefs is associated with one or more of these goals on an operating level. On a more fundamental level, these goals boil down to two fundamental needs: the needs for acceptance and control.

The Need for Acceptance

I define comfort as self-acceptance. When we finally learn that self-care begins and ends with ourselves, we no longer demand sustenance and happiness from others.
Jennifer Louden

Humans are social animals. We need to be accepted by other humans because there is safety in numbers, as we discussed this in Chapters 2 and 3. Our need to be accepted can lead us to adopt negative beliefs and to take actions that limit our own individuality and priorities. The need for acceptance can cause people to do al kinds of negative things, such as become anorexic or bulimic, join cults, suppress their sexuality, or live under otherwise intolerable circumstances. How many people do you know who were pressured to have the right friends, go to the right school, or get the right job? To what extent do your own worries come from feeling like you have to keep up with the mythical Joneses? Before you feel too sorry for yourself, consider that the Joneses are probably in the same boat. After all, everyone knows someone who seems to be better off than they are.

The need for acceptance is hard-wired into our brains. We can never get rid of it entirely. But we can learn to harness this need. It starts with learning to unconditionally accept ourselves for who we are. It starts by never again saying, "I am ____," even if the word you'd use is positive. The truth is that every conceivable word you could use is nothing but a subjective label that has nothing whatsoever to do with who you really are. Learn instead to simply say, "I am!" No labels. No descriptions. No qualifiers. No judgments. Nothing. Just you. You are, period. When you accept yourself unconditionally, then and only then will you be ready for true acceptance from others. The strategies outlined in the rest of this book will help you get there.

Of course there is no formula for success except perhaps an unconditional acceptance of life and what it brings.
Arthur Rubinstein

One key side effect of learning to accept yourself unconditionally is that you will no longer feel the need to jump through any hoops to be accepted by others. Another side effect is that learning to accept yourself teaches you how to accept others. You need not like or get along with anyone. Acceptance means just that: You recognize their fundamental value as human beings no matter what. Together, these two side effects mean that you will gravitate toward people who will accept you unconditionally, and vice-versa. In short, you will have all the acceptance you need without any of the pretenses that stemmed from—and helped fuel—your negative core beliefs.

The Need for Control

If everything seems under control, you're not going fast enough.
Mario Andretti

We all need to feel like we're in control of our lives and our environments. We like to believe that there is such a thing as security and that we can bend the forces of nature to our will. On the other hand. losing control is a very scary feeling. I can't imagine a worse feeling for a prey animal than hopelessness.

The standard Copenhagen interpretation of quantum physics implies that the observer creates her or his personal reality by "collapsing" all possibilities into a concrete result. Imagine a subatomic particle like an electron. According to quantum physics, that particle has no specific location. It cannot even be said to exist as anything other than a *probability wave* (called the *quantum waveform)* that says, in essence, "When observed, the electron will have an X% chance of being at Location A, a Y% chance of being at Position B, etc." This means that the electron will appear at a specific location when you look at it, only to vanish into another wave when you look away. The location where the particle appears depends on the probability of it appearing at any possible location. For example, if the probability of the electron being in a certain position is 75% for Position A and 25% for Position B, then it will appear in Position A three out of every four times you observe it, on average. It will appear at position B once for every four times you observe it, on average.

When you look at an ordinary object, you are "collapsing" the positions of all of that object's particles at once. Extending this logic reveals the startling fact that no object can be said to exist when you are not looking at it! Even so, unobserved objects will continue to behave according to quantum laws, which makes their reappearance when you look again all but guaranteed. Yes, there is a chance that my computer will spontaneously jump to Pluto; however, the odds of this happening are vanishingly small.

A reasonable probability is the only certainty.
E.W. Howe

More than one New Age guru or expert claims that quantum physics means you can have anything you want, according to the "Law of Attraction." This much is true, at least on a purely technical level. What that guru or expert won't tell you is the odds of that actually happening. Say you want to manifest $1,000,000 in $100 bills. You would need 10,000 of these bills to add up to the full million dollars. Let's say that each bill consists of 100 atoms. (The actual number is well over a million trillion.) Let's further say that the odds of any one of those atoms jumping to a given location on command is 50%, or 1 in 2. (The actual odds are many orders of magnitude less.) The odds of getting just 2 atoms together in front of you is therefore 0.5x0.5=.25, or 1 in 4. The odds of getting 3 atoms together is 0.5x0.5x0.5=0.125, or 1 in 8. Your odds of getting a single hundred dollar bill in front of you are thus approximately 1 in 10^{30}. That still leaves you $999,900 shy.

Do not expect to arrive at certainty in every subject which you pursue. There are a hundred things wherein we mortals... must be content with probability, where our best light and reasoning will reach no farther.
Isaac Watts

It gets worse: The entire age of the Universe is only about 4.7×10^{17} seconds. If you started making one attempt per second to manifest a single hundred-dollar bill at the moment of the Big Bang, you would need to wait 2,127,660,000,000 times the current age of the universe to be successful, on average. You would then need to do this 9,999 more times to finish manifesting your million dollars. If this example does not convince you that the "Law of Attraction" is worse than useless, nothing will.

All of this math is a very long way of saying that yes, you can indeed have anything you want, and the "Law of Attraction" is very real—provided you'll be around long enough to see your desired result! You are far better off playing the lottery where the odds are only about 1 in 1.2×10^{8}, which is positively reasonable by comparison. No, I am not suggesting that you start playing the lottery! Your odds of winning the super jackpot are mind-numbingly small, but they are still many orders of magnitude better than your odds of creating $100 out of thin air!

God grant me the serenity to accept the things I cannot change; courage to change the things I can; and wisdom to know the difference.
The Serenity Prayer

You can use your resources to procure other resources, such as buying things or persuading others to see your point of view. You can choose how to interpret things that happen to you. The bottom line is you can indeed control yourself. You can also wield some external influence. But when it comes right down to it, you have far less control over your environment than you think you do. The trick is to recognize what you can and cannot control. Which brings us to the topic of...

The secret of concentration is the secret of self-discovery. You reach inside yourself to discover your personal resources, and what it takes to match them to the challenge.
Arnold Palmer

Resourcefulness

The previous chapters have been all about getting rid of your negative mental programming by simply letting it go. Negative beliefs drain your physical, emotional, and spiritual resources. Breaking down the word *resource* gives you *reserve* and *source.* A resource is therefore a quantity, or reserve, of source. Source, of course, is defined as "the beginning or place of origin." Whenever you have a resource, you therefore have a piece of the beginning or origin. Follow this logic, and you'll soon understand that every resource is literally a piece of the origin of the Universe. What is the origin? For reasons that it will take me all 1,000 pages of *The Divine Savage* to explain, the origin is love. When you have a resource, any resource, you have love. When you are in a resourceful emotional state, you are in a state of being full of source, or full of love. Love is the highest possible emotion and is, I believe, the source and reason for all things. Being full of love connects with the divine, that which is beyond evolution.

> *Love is an exploding cigar we willingly smoke.*
> Linda Barry

From everything we have discussed so far, it should be obvious that the opposite of love is not hate but fear. Fear is an evolved emotion that developed to aid individual survival and is especially strong in prey animals, including humans. As I explain in *The Divine Savage*, the mere existence of an individual organism implies at least some separation from the divine source, which implies a capacity for fear. All intermediate emotions fall into a continuum between abject fear (hopelessness) and abject love (peace). Let's take a look at some of these emotions:

Hopelessness

Hopelessness is the "absolute zero" of the emotional scale. In fact, it's almost an emotionless state to all outside appearances. Hopeless people have lost all drive and motivation. They feel stuck and unable to take any action to change things. Being around a hopeless person is an intensely draining experience for anyone. The hopeless person finds her or himself increasingly isolated and bereft of apparent options. S/he is cornered, however figuratively, and has lost the ability or will to care. S/he is beyond fear.

> *The body is very loyal. It will even die for us if we tell it to.*
> Jim Britt

All of us experience hopelessness from time to time. Most of us can snap out of it relatively quickly, but some people get caught in an ever-descending spiral. In this state, one often feels like life has punched them in the gut. The underlying issues have probably been building for some time. On the plus side, this means that those issues can be peeled back one by one.

Sadness

We ask God to forgive us for our evil thoughts and evil temper, but rarely, if ever ask Him to forgive us for our sadness.
R.W. Dale

Sad people may feel lonely, depressed, or both, but they haven't thrown in the towel. Sadness is an expression of fear. Grief over the death of a loved one is an expression of the fear over one's own mortality and the possibility that death is the end of all existence. Sadness over something that happened to you directly is an expression of fear of the consequences. One can think of sadness as fear of something that has already happened or that could/will happen in the future. Another way to put it is that sadness is a time-displaced form of fear.

Fear

The greatest mistake you can make in life is to be continually fearing you will make one.
Elbert Hubbard

Fear is defined as, "agitation and anxiety caused by a present or imminent threat." Unlike sadness, fear exists in the now. What is our ultimate fear? Being killed and eaten. An unarmed human confronted by a predator animal has fairly long odds against surviving the encounter.

Anger

The Chinese use two brush strokes to write the word 'crisis.' One brush stroke stands for danger; the other for opportunity. In a crisis, be aware of the danger—but recognize the opportunity.
John F. Kennedy

Anger seems to come from the need for control but actually stems from an extreme need for acceptance. An angry person wants the object of that anger to accept her or his point of view. An angry person needs to be "right" (accepted), which spurs a corresponding need for control. It is no coincidence that this need for acceptance shows itself as a power/control play. The underlying fear is of a loss of control that causes a loss of acceptance. A key difference between anger and outright fear is that the angry person has much better odds of getting her or his way, which means that s/he has a much lower chance of getting killed and eaten. Thus, an angry person thinks s/he has a chance to "win" the situation.

Pride

The charity that hastens to proclaim its good deeds, ceases to be charity, and is only pride and ostentation.
William Hutton

Pride is another fear-based emotion. Think about the so-called "high achievers" you know. How many of them seem judgmental, selfish, snobbish, rigid, narrow-minded, opinionated, set in their ways, vain, and/or all-knowing? How many of them are constantly blowing their own horns, either overtly or more subtly? People like this are self-described paragons. They have a desperate need for acceptance, which often comes across as controlling.

Courage

Courage is the emotion that lets you act in spite of fear. Standing firm and resolute in the face of danger is a resourceful, love-based thing to do. Being truly brave means loving yourself and/or someone else enough to take a chance. Courage takes many forms but always involves setting aside some of the need for acceptance and/or control. Setting aside any portion of your two fundamental needs can only come from a place of security and confidence... from source... from love.

Courage is the ladder on which all the other virtues mount.
Clare Booth Luce

Acceptance

Acceptance does courage one better. You are willing to set aside some of your need for acceptance and/or control without feeling the need to take action or stand firm. When you accept a situation, you adopt an attitude of, "whatever will be, will be." You don't have to like the situation to accept it.

Acceptance is not submission; it is acknowledgment of the facts of a situation. Then deciding what you're going to do about it.
Kathleen Casey Theisen

Peace

Peace is the ultimate state of resourcefulness or love. When you are peace, you don't have any need for acceptance or control. You face the possible consequences with total equanimity. There is no fear whatsoever. There is only source. There is only love.

Nothing is more conducive to peace of mind than not having any opinions at all.
Georg C. Lichtenberg

The Art of Letting Go

I used the term "absolute zero" to describe the emotion of hopelessness. All emotions involve energy. Break down the word "emotion" and you get "energy in motion." Increasing the energy of a photon causes its frequency to rise from the radio and microwave bands (at the low end) through the infrared, visible light, ultraviolet, X-ray, and gamma bands. Hopeless people are like a photon at the lowest end of the spectrum. People who are at peace are like photons at the highest ends of that same spectrum. The brighter the light, the more photons are traveling. The stronger the emotion, the stronger the underlying beliefs. Just as photons color everything we see, your beliefs color everything you perceive with emotion. Your beliefs emit the "photons" of emotion.

Emotions are our response to the things that happen in our lives.Again, the things themselves are neither good nor bad; they just are. If you have negative core beliefs, you will project a low-frequency emotion that will remove you from source. Letting go of a fear-based emotion literally dumps the energy that is being spent on that emotion and frees it up for anything you want, including a love- or source-based emotion.

This has nothing to do with thinking positive or putting on a happy face. This has nothing to do with affirmations, even though affirmations can help you prepare for the "moment of truth" when you experience an emotion or (even better) recognize an event and choose how to label it. This has everything to do with releasing the energy that is generated by, and that reinforces, your negative beliefs.

To be one's self, and unafraid whether right or wrong, is more admirable than the easy cowardice of surrender to conformity.
Irving Wallace

Letting go means releasing yourself from whatever flotsam you're clinging to. Jim Britt tells the story of a woman at one of his seminars who told him that she could not life her ideal life because her father had abused her. Jim asked her where her father is and learned that he had been dead for several years. Jim then asked, "So who is abusing you now?" It took the woman a few minutes to realize that she was abusing herself. This realization helped her let go of her negative core beliefs about her father and replace them with newer, more positive programming. Her process was much the same as the processes you experienced in the previous chapters.

Letting go of negative emotions before they take hold is crucial in the same way that putting out a fire before it spreads is crucial. Letting go must happen in the present, which means that learning to let go means learning to live in the now. Now is where the action is and where the solution is. Letting go is resourceful. It is living in a state of love, of yourself and the entire Universe. It is acting from love. Acting from love affects you and all those around you.

Of course, the best way to put out a fire is to prevent it from happening in the first place. The best way to let go of a negative emotion is to not experience that emotion in the first place. This requires you to realize some very fundamental truths and adopt a world view that can be summarized as...

Non-Attachment

Nothing is permanent. Everything changes. This change is usually toward decay and disorder thanks to the Second Law of Thermodynamics, which says that the total amount of *entropy*, or disorder, must always increase. It is much easier to lose money than to make it. You cannot un-spill a cup of coffee. Bodies age and eventually die. There are no guarantees in life. Let me give you a personal example: My partner Jennifer left for the gym a couple hours ago. The simple truth is that she may be dead right now. She may have decided to terminate our relationship. I have no reason to think that either of these things has happened but that is no guarantee that they haven't. Anything and everything in my life that I am not directly observing right now as I type this could have changed. The only guarantee I have is that nothing in my life is permanent, not even my life itself. You and everyone else are in the same boat. Everything that we think of as good in our lives will vanish at some point. On the other hand, so will everything we think of as bad. It's a zero-sum game, especially since there is no such thing as intrinsic good or bad.

When you relinquish the desire to control your future, you can have more happiness.
Nicole Kidman

Attaching or identifying yourself with anything in your life by saying, "I am ____" is therefore a recipe for eventual disaster. A friend of mine whose husband committed suicide spent a year "finding a new normal" because she identified a huge part of herself as, "I am wife." Losing her husband caused her to lose a huge part of her identity—an incredibly traumatic experience.

Realizing the impermanence of everything is incredibly liberating. The woman in Jim Britt's seminar realized the impermanence of her father's abuse. I love Jennifer with all my heart and I choose to see the three-plus years I've had with her as some of the best in my life. Nothing and nobody can ever take my time with her away from me, nor can anything diminish the huge blessings I've received from our relationship. When the time inevitably comes for our relationship to end through death or dissolution, I will be happy and grateful for the time we had. Should something happen to my son, I will forever be blessed for having been his father. Everything in life is like a piece of cake that you are eating. You know full well that the cake will eventually be finished and yet you free yourself to enjoy it while it lasts with nary a moment of grief or regret or "finding yourself" once the plate is empty. The ironic part is that the cake ends up being far more "you" than most things in your life because its atoms and molecules become part of your body.

On action alone be thy interest, never on its fruits. Let not the fruits of action be thy motive, nor be thy attachment to inaction.
Bhagavad Gita

Being unattached also removes all expectations. Expectations are a zero-sum game at best. If I give you something and expect something from you in return, then the best I can do is break even. If you don't fulfill your end of the bargain then I lose. On the other hand, if I give you something with no expectations, then the worst I can do is break even. It's a no-lose proposition for all concerned. My loving Jennifer with no attachments allows me love her freely to base my desire to be with her on my love instead of the other way around. Loving her without expectations frees me to receive so much more without her having to give anything more... because I don't expect anything.

You have very little control over the many events and people in your life, which means that you have very little real power to meet your needs for acceptance and control. It's OK to want things and to plan and move toward goals. It's OK to want love and meaningful relationships. Just don't attach yourself to any particular outcome, and don't expect anything in particular to happen. If letting go of negative emotions is putting out a burning fire, then non-attachment is preventing the fire from happening in the first place.

Wisdom is what's left after we've run out of personal opinions.
Cullen Hightower

There is a Chinese proverb about a farmer who had a prize stallion. One day, the stallion escaped and ran off. His neighbors offered their condolences for this misfortune, to which the farmer replied, "Good or bad? Who knows?" Three days later, the stallion returned leading a herd of wild horses. The neighbors congratulated the farmer his good fortune, to which the farmer replied, "Good or bad, who knows?" The farmer's son tried to ride one of the horses but fell off and broke his leg. Again the neighbors offered their condolences, and again the farmer replied, "Good or bad, who knows?" Several days later, the army came through conscripting able-bodied men to go off and fight. The farmer's son was exempted because of his injury. And so it went. So it goes in all of our lives. The only question is whether or not we can remain unattached enough to see the neutrality and opportunity in everything that can possibly happen to us.

Believe it or not, non-attachment is an expression of control. You constantly remind yourself not to attach yourself, which means you exercise control over your identity. Non-attachment is an expression for the need for acceptance because you will accept whatever comes and will also seek out people who will accept you are you are. Developing non-attachment requires...

Conscious Choice

Because you are in control of your life. Don't ever forget that. You are what you are because of the conscious and subconscious choices you have made.
Barbara Hall

As we discussed in Chapter 4 and elsewhere, your emotions are not the cause but the result of the process of building your own personal reality. Most of us choose how to interpret the events in our lives subconsciously, meaning that you have already labeled them as good and bad by the time you experience the emotion. The emotion is therefore already positive or negative by the time you feel it. It is either drawing you closer to—or further away from—source, or love.

Letting go means recognizing the emotion when you experience it and choosing to stop having that emotion. This happens when your logic has kicked in to rationalize or explain the emotion. Letting go of the emotion means that you only experience it for a few seconds (at best) or that its intensity and/or duration is reduced. This is a good thing!

Letting go of your attachments and expectations helps prevent negative emotions from happening in the first place. This is ultimately where you want to be. This requires conscious mental processing to accomplish.

The other aspect of conscious choice involves living your life on purpose. How many choices do make by default every day? If you are in a relationship, then you continue on in that relationship. If you have a job, then you keep going to work every day. Many times this happens unconsciously; you go about your routine because it is routine. This is the equivalent of staying in your burrow just because you've always stayed in your burrow. Live like this and you will endure a lot before the pain of staying becomes greater than the pain of leaving, or until something happens over which you have no control (which is just about anything). This is how most people live their lives.

On the other hand, imagine making new choices every day. Imagine waking up every morning and choosing to be with—to really be with—your significant other? How would making this choice influence your behavior toward that person or how you interpret their behavior? Every morning, as I wake up, I ask myself whether I want to be with Jennifer today, just today, because that is all I control. Making that choice gives me the gold standard by which to evaluate my behavior toward her. Every time we interact, I can ask myself: Is how I am interacting with her supporting or detracting from my decision?" If the former, fine. If the latter, then I have the ability to stop myself in midstream and choose to behave differently. Choice is power. By choosing to be with Jennifer, I take full responsibility for everything I bring to the relationship. My constantly renewing my decision to be with her constantly reminds me of my responsibility to her and to our relationship. It also gives me the power to say no, that I no longer want to be with her. Knowing that I am with her by choice and reserving the power to change my choice means that I can be fully with her without reservations. Giving myself the freedom to leave gives me the freedom to stay. The same logic holds for everything in my life. Giving myself the power to choose how I live my life gives me the ability to do almost anything I like.

How do you get to the point where you start automatically making conscious choices? You get there by developing a mental "third eye" that watches over everything that happens to you and intervenes whenever something happens. This is something that I call...

The Conscious Interrupt

The world is moving so fast these days that the man who says it can't be done is generally interrupted by someone doing it.
Harry Emerson Fosdick

I spent several months feeling like a total failure as my marriage to Robyn ended and my finances collapsed around me. Here I was, 40 years old, with nothing to show for my time on Earth. Jennifer repeatedly told me that I was rich because I had so much going for me, but I couldn't bring myself to believe her. I alternated between hopelessness, sadness, fear, and anger. Most of all, I felt like a fraud. After all, hadn't I just finished writing a book (the first edition of *The Enlightened Savage*) that claimed to tell people how to avoid the very kinds of problems I was facing? In fairness, the processes I included in the first edition of this book did work quote well. I often marveled at how calm and collected I was in the face of the worst set of crises I've faced to date in my life. But they hadn't prevented this calamity from befalling me.

As I researched *The Divine Savage* and worked with Jim Britt on several projects, it gradually occurred to me that the science I used is perfectly valid. The problem with the former edition of this book was not that it was wrong, but that it only painted part of the total picture. I was doing a great job of replacing my old beliefs, but was not following up by exercising conscious control over my responses to events. I therefore resolved to become aware of everything that happened, no matter how minor. I trained myself to recognize when a situation with possible emotional consequences was taking place.

> *That consciousness is everything and that all things begin with a thought. That we are responsible for our own fate, we reap what we sow, we get what we give, we pull in what we put out. I know these things for sure.*
> Madonna

Then, I trained myself to choose how to respond. Letting go of attachments to labels and the idea of intrinsic good and bad was a huge help. Fully opening myself to the impermanence of this lifetime and internalizing the lessons of the proverb I shared above opened me up to see a wider truth. Yes, my marriage was over, but Robyn and I are better friends who love each other more and who get along better than we have at any time in our past. Jennifer didn't care about my finances; she loves me for me. My son is one of the smartest, funniest, kindest, most well-adjusted kids I know. Friends and relatives stepped forward to help me in my time of challenge. Suddenly, my financial situation didn't seem so bad. In fact, I'm thankful for it! I've been calmer, mellower, and happier than ever. Going forward, loving the people in my life without being attached while consciously interrupting myself before I can form an opinion are ends unto themselves that will eventually lead to even greater material abundance. Most of the wealthiest people I know went through one or more bankruptcies before finally obtaining financial freedom.

So what exactly is this conscious interrupt? It is exactly what the name implies: It is conscious, meaning that you are aware of it at all times. It's like a little person sitting on your shoulder

Everyone needs a strong sense of self. It is our base of operations for everything that we do in life.
Julia T. Alvarez

watching everything going on in your life like a hawk. This little person never sleeps. He is always vigilant. And when some event happens, he interrupts your reaction in mid-stream. This interruption is like a ringing telephone that jars you out of a daydream. The whole idea is to replace the choices you already make unconsciously with similar choices that you can make consciously.

Questions

This chapter helped you set up a basic mental framework that will help you get the most of the strategies we'll be covering the remaining chapters. This framework consists of understanding your fundamental needs, resourcefulness, letting go, non-attachment, conscious choice, and developing a "conscious interrupt" that watches over you and intervenes to give you the opportunity to make conscious decisions.

Write your answers to the following questions:

- **What are the things you feel you need to control in your life?**
- **How many of these things do you actually have control over?**
- **Who do you feel you need to be accepted by?**
- **Why do you feel that you need these people to accept you?**
- **Can you imagine people who would accept you more than the people who are currently in our life? If so, what is keeping you from finding these people?**
- **Do some or all of the people near you accept you conditionally based on expectations or beliefs about who you are or about "what's right?"**
- **Do you accept yourself unconditionally? If not, why not? How can you learn to accept yourself for who you are without any conditions or expectations?**
- **On a scale of 0-10 with 0 being never and 10 being always, how often do you feel hopeless?**
- **What causes you to feel hopeless?**
- **How can you feel less hopeless?**

- **On a scale of 0-10 with 0 being never and 10 being always, how often do you feel sad?**
- **What causes you to feel sad?**
- **How can you feel less sad?**
- **On a scale of 0-10 with 0 being never and 10 being always, how often do you feel afraid?**
- **What causes you to feel afraid**
- **How can you feel less afraid?**
- **On a scale of 0-10 with 0 being never and 10 being always, how often do you feel angry?**
- **What causes you to feel angry?**
- **How can you feel less angry?**
- **On a scale of 0-10 with 0 being never and 10 being always, how often do you feel prideful?**
- **What causes you to feel prideful?**
- **How can you feel less prideful?**
- **On a scale of 0-10 with 0 being never and 10 being always, how often do you feel brave?**
- **What causes you to feel brave?**
- **How can you feel more brave?**
- **On a scale of 0-10 with 0 being never and 10 being always, how often do you feel accepting?**
- **What causes you to feel accepting?**
- **How can you feel more accepting?**

- **On a scale of 0-10 with 0 being never and 10 being always, how often do you feel at peace?**

- **What causes you to feel at peace?**

- **How can you feel more at peace?**

- **What things do you feel you need to let go of in your life? Provide as much detail as possible.**

- **What things do you feel attached to and how can you learn to become unattached from them?**

- **On a scale of 1 to 10 with 1 bering rarely and 10 being always, how often do you make conscious decisions about how to interpret the events in your life, no matter how minor they may seem? ______________**

- **f you answered anything less than an 8 or a 9, how can you develop your ability to be conscious to that level? Be as specific as possible.**

- **Do you have a "conscious interrupt"? If not, how can you develop one?**

- **How else do you think you can use the framework presented in this chapter to help you make use of the strategies that you're about to learn?**

Chapter 16

Maintenance and Growth

The great awareness comes slowly, piece by piece. The path of spiritual growth is a path of lifelong learning. The experience of spiritual power is basically a joyful one.
M. Scott Peck

By this time, your excitement might be tempered by the need for a little more encouragement and direction. You may even be wondering if the changes you just experienced in the previous chapters are real, and praying they'll last. Those are valid concerns, because your new beliefs and expectations are like newborn children that need your love and nurturing in order to grow strong and healthy. Your old programming has been with you your entire life, while your new beliefs and expectations are only minutes, hours, or days old. You've come a long way but will slide right back without ongoing care, y. The processes you just experienced function like surgery that mends a physical ailment. Just like surgery patients require ongoing therapy to fully heal, you need ongoing strategies to bolster and reaffirm your new programming. The previous chapter gave you the framework for using the strategies you are about to learn.

This chapter contains fourteen great success strategies. Use them like physical therapy for your recently operated-on beliefs, and those beliefs will grow big and strong. Changing the programming was easy. The good news is that maintaining your new programming is just as easy. In fact, it's even easier because all you are doing is mastering that which you have already built.

You know that your brain is designed to build walls around your safe little worlds, and to resist any and all attempts at change. You also just got done knocking down the walls surrounding your old beliefs. The next step is to build new walls around your new beliefs.

> *The quality of our expectations determines the quality of our actions.*
> Andre Godin

Until very recently, you focused on things that you interpreted as negative and concentrated on the worst-case scenario. You weren't alone; most people remain stuck in a pattern of interpreting things negatively because humans are prey animals. Here are some examples of typical prey thinking:

- I won't get promoted, and this is bad.
- No one will buy my product, and this is bad.
- I'll get ridiculed, and this is bad.
- I'll never be happy, and this is bad.
- That's life, and this is bad.
- I'm powerless, and this is bad.

The list goes on forever. You know exactly what I'm talking about, because you were living that personal reality up until just a little while ago. You now know that none of the things I listed above are negative, nor is anything positive. You make them positive or negative with your choices, which means that you could choose to see any of the above as a blessing.

What Really Changed?

> *Who we are never changes. Who we think we are does.*
> Mary S. Almanac

It may seem like I'm telling you to fight millions of years of evolution and ignore your brain's survival instincts. You may also perceive another gaping flaw in my logic: If we were prisoners of our old beliefs, didn't we just become prisoners to those new beliefs to the same extent that we were once prisoners of our earlier beliefs?

Absolutely! In fact, this is the point of the entire book.

Breaking through the walls of your old beliefs lets you step into a new room to start building new walls. Your brain is still wired for survival and will still resist all attempts at change, lest it be killed and eaten. So what's changed?

Not much. In fact, the only thing we've done is changed your definition of survival. Instead of defining survival and success based on negative or destructive beliefs, you'll be defining them based on positive and constructive beliefs. Your brain will learn to resist doing anything non-constructive in the same way it learned to avoid the success you want and deserve.

Taking Advantage of Prey Instincts

Follow your instincts. That's where true wisdom manifests itself.
Oprah Winfrey

You will act out of the same primal motives, but with vastly different results. You will act in spite of your old fears and doubts, not because of them. In other words, you will be afraid of your negative patterns in the same way that you used to fear positive patterns. This is what I mean when I say that all you are doing throughout this entire book is learning how to use your built-in prey instincts to your advantage. you are literally exchanging your physical addiction the emotional chemicals associated with not living your dharma for the physical addiction to the emotional chemicals that motivate you to live your ideal life. Afraid to make sales calls? You'll relish them. Don't think you deserve happiness? You won't settle for anything less than bliss. Think work equals struggle? You'll start producing more results with less effort than ever before. You see how this works? And all along, you will be making choices about to interpret each and every little thing that happens in your life. Develop the ability to make conscious choices using the "conscious interrupt" I mentioned in the previous chapter and nothing bad ever need happen to you again... literally.

As you can see, the whole idea is not to go against your brain, but to turn how you brain works to your advantage. Think about your car: You can't affect the engine's inner workings, but you can use that engine to take you anywhere you want to go. Same with your brain, except that the only traffic cops in your head are the ones you put in your own way.

There's another key difference: Your new programming makes the walls surrounding your new beliefs transparent. Use the success strategies I'm about to share with you, and you'll find yourself consciously monitoring and adjusting your beliefs, and listening to what your brain is saying.

Success Strategies

These strategies were developed by Caterina Rando for her amazing *Success With Ease* program.

Whenever something potentially negative pops up, you'll be able to recognize and neutralize it, then choose to see the good in that event—a lifelong process of debugging and improving your programming instead of running whatever loose code you have in your head. Here's how you do it:

Most people give up just when they're about to achieve success. They quit on the one yard line. They give up at the last minute of the game, one foot from a winning touchdown.
H. Ross Perot

Choose Joy

Life is a series of choices. Yes, challenges do occur and yes, they sometimes happen to the most positive among us. I believe that each challenge occurs because the Universe is trying to teach us something that we can use to further our own growth and success. Never forget that the Universe loves you too much to let you fail and will never give you more than you can handle.

Joy runs deeper than despair.
Corrie Ten Boom

Remember the old saying: If life hands you lemons, make lemonade. Look for the joy in every situation. It may be difficult sometimes, but if you force yourself to consciously look for and choose joy, then joy will come. What you sow, you reap.

Live on Purpose

Each of us is on this planet for a reason. My mission is to help people change their inner programming. What about you? What makes you passionate? One man I talked to was born to paint. No wonder he was having trouble finding work as a graphic designer. Find your life's mission and live it. We've discussed dharma. Follow your dharma, and your life will be rich beyond your wildest dreams. There is no shortcut or substitute for living on purpose.

Many people have the wrong idea of what constitutes true happiness. It is not attained through self-gratification, but through fidelity to a worthy purpose.
Helen Keller

Acknowledge Others Often

Each of us depends on many other people for just about everything in life, such as the people who made my clothes, the people who made the computer capturing the words and format of this book, or the farmers who grow my food. Same with you. You are not an island. No one is. From the maid

If you wish your merit to be known, acknowledge that of other people.
Anonymous

who cleans your home to the waiter who pours your water or the spouse who goes to the office all day, take the time to acknowledge their presence and contributions to your life, whether large or small, direct or indirect.

Don't go overboard. A simple "thank you," a fair tip, a smile, common courtesy—That's all. If someone treats you badly, respond with even greater courtesy and kindness. My friend Joe once gave me a great piece of wisdom: People never act against anyone else, just for themselves—and they usually do so because of their own self-destructive beliefs. Ponder that a while. I'm positive you'll start seeing people in a whole new light. Treat others as your new successful beliefs want them to treat you. It's that simple and that easy.

Ask for What You Want

We find what we expect to find, and we receive what we ask for.
Elbert Hubbard

My musician friend wanted to rebuild her career. I told her to contact everyone she knows in the industry, explain her situation and desire, and ask for help. No pride, no long explanations, no negative self-talk, no obfuscation, nothing. Bare your soul and ask for what you want... and be completely unattached to whether you get it or not. Getting what you seek can entail unforeseen challenges. Not getting what you seek can be a huge blessing in disguise.

Think about it. You want things and probably know people in a position to help you. Ask! My friend contacted a producer who agreed to help her at his expense. Jay Conrad Levinson had no idea I existed until I emailed him asking him to review my book marketing programs for independent authors. The worst he could have said is 'no', but he could never have said yes until I asked. Jim Britt had no idea I was working on this book and its sequels until I asked him for help getting them out to the world. So ask! Chances are, someone wants what you have to offer. I've asked for help on many occasions. So can you.

Be Willing to Be Uncomfortable

To be uncertain is to be uncomfortable, but to be certain is to be ridiculous.
Chinese proverb

I've said it before, and I'll say it again: Your brain will throw up all kinds of obstacles as you work to develop your non-attachment and your conscious interrupt. Even once you have these finely tuned, you will still find yourself slipping from time to time. Here's an email I once received. I quote it verbatim:

> *People have to work through things in their own way. Your advice is like telling someone with major depression, "Oh, cheer up." I understand that you have the drive to change your reality or at least your vision of it. That's a wonderful thing. But for a lot of people, it comes down to DNA and lots of therapy. Some have to take that journey one step at a time to build a solid foundation. I had to learn this the hard way, and no amount of sheer will and positive thinking changed my reality.*
>
> *Obviously, I had yet to reach that higher level of consciousness that would allow me to defy basic physics or "rise above" the things that had happened to me, but this is where most of us live. Granted, most of us can incorporate a lot of your message positively into our lives, but even so, we still have a lot of hard work to do. Even though life is a gift, sometimes it does suck.*

This email is a classic example of someone who is unwilling to experience discomfort by leaving the comfort zone. It is also an example of the power our beliefs hold over us, because the author is clearly unwilling to even explore any alternatives to his struggle-based personal reality. Read the letter carefully. Do parts of it sound familiar? After watching this far, and after changing your own programming, how would you reply?

Explore New Possibilities

Having defined your life's goals, how can you use them to get the kind of success you want, instead of the kind if success that takes you further from what you want? If you're a painter by passion and calling, how can you turn your painting into financial success and personal freedom? Once you've found your life's mission, look for new possibilities every day. Think outside the box; try unconventional approaches. Put this fresh spin on your chosen field. Combine this strategy with just

> *God... created a number of possibilities in case some of his prototypes failed—that is the meaning of evolution.*
> Graham Greene

about any other success strategy, and you'll be that much closer to achieving your goals.

Maintain a Positive Disposition

Men of the noblest dispositions think themselves happiest when others share their happiness with them.
Barry Duncan

Everyone has tough days. Remember this: The Universe, or God, or whatever you believe in, loves you, and is trying to give you a miracle every day. Don't look that gift horse in the mouth. Each day contains new opportunities wrapped in new challenges. The greater the challenge, the greater the opportunity. Health problems, an accident, loss of a loved one, etc. can all be tragic if you choose to see them as such, and you may choose to grieve.

Just never lose sight of the gift buried in the turmoil, and never ever forget that no event is ever positive or negative until you make it so by will or by default. Developing non-attachment prepares you for the eventual end of all things in this lifetime and frees you to enjoy them in the present without conditions. Your conscious interrupt can then help you find the gift that always exists under absolutely any circumstances. Finding that gift honors its source and makes it a contributor to our lives instead of a drain. When dealing with customers, investors, suppliers, employees, and the like, always remember the four C's:

- Cool
- Calm
- Collected
- Confident

I once frequented a great little breakfast joint, and knew the owners on a first-name basis. One day, Robyn and I showed up for a late brunch. It was obvious that we'd walked into the middle of a fight. The palpable hostility in the air made that meal so uncomfortable that we never returned. The lesson is obvious.

Take Small Actions...

Happiness is not something ready made. It comes from your own actions.
Dalai Lama

...towards a huge outcome. Achieving your goals might seem light years away. Staring up at a tall mountain from the bottom can be a daunting experience if you intend to climb it. If you find yourself focusing on the enormity of the road ahead, that is your brain trying to hold you back.

Thanks for the step up! Don't worry about the mountain. It will still be there. Focus on climbing the first mile, the first half mile, or even that first single, tentative step. That's not too hard, is it? You can walk or climb a short distance. Once you reach that goal, pick another one a little further away and set off. If you find yourself contemplating the mountain, thanks for the step up! Force yourself to focus on your next milestone. Celebrate every tiny step as the achievement it is. Don't quantify or reduce it. Just enjoy it.

Bit by bit, mile by mile, you'll climb higher. At some point, you'll pause to take in the scenery and be astonished by how far you've come. And the scenery just keeps getting better! How great is that? Worried that the higher you climb the farther you might fall? Hello, climber: This is your brain speaking! What are you going to say? "Gee, you're right"? Or, "Thanks for the step up?" Besides: If you use the proper climbing equipment, the most you can fall is a few feet—hardly fatal. From there, you can pick an easier route and keep going. These ropes are nothing more than very carefully laid plans that include provisions for contingencies. Yes, setbacks may occur. You have the power to minimize them or to be overwhelmed. What's it going to be?

Openly Express Your Gratitude

Whether it's the person who gives you your first big break or the clerk who rings up your groceries, your life is better because of these people. It doesn't stop there. You would not be here but for God, the Universe, or whatever driving force you believe in. Every day above ground is a new day with new blessings and ways to grow. Accept the gifts that are coming to you without preconditions, attachments, or judgment and be thankful! Above all, don't forget to thank yourself.

There is a calmness to a life lived in gratitude, a quiet joy.
Ralph H. Blum

If "I already know that" are the most destructive words in any language, then "thank you" are the two most powerfully uplifting words in any language, provided they are said simply and sincerely. You need not be effusive; in fact, gushing diminishes your thanks. Thank you. Two words with unimaginable power. A warm smile. A nod. A helping hand. The little things. Those have meaning.

Never, ever say "I owe you". The person helping you wouldn't have done so if it wasn't in their best interests. Even the most altruistic acts have self-interest at heart. Don't worry about repayment. You either have an agreement in place or some situation will arise where you can repay the kindness. Maybe not to the person who gave it to you, but to someone else. And that is perfectly valid.

Have Some Fun

Work for the fun of it, and the money will arrive some day.
Ronnie Milsap

All work and no play will make you one dull, stressed-out individual. Work your fingers to the bone, and you'll have bony fingers. You get the idea. The whole point of this exercise is to succeed with ease. Never forget that work means results, not effort. Why struggle when an easier solution may be available?

If you're succeeding with ease, then you'll have more time to do other things. What should you do with all this free time? Recharge your batteries! Take that trip you've always wanted to take. Go bungee jumping. In fact, why not combine this with the discomfort tool and have an adventure that challenges your comfort zone? Whether it's zipping along a line between two trees, rock climbing, skydiving, scuba diving, mountain biking, or any other adventure you can imagine, the best way to expand your boundaries is to push them—within reason, of course. Always take appropriate precautions, and be very sure that you always remember the distinction between adventure and foolhardiness. Gravity is neither good, bad, nor forgiving.

Smile, Laugh, and Love More

Let no one ever come to you without leaving better and happier. Be the living expression of God's kindness: kindness in your face, kindness in your eyes, kindness in your smile.
Mother Teresa

Whatever you put energy into grows! If you don't have all the happiness, laughter, and love you want in your life, put that energy out there. It will come back a thousand fold, my friends.

The energy you receive will lift your spirits, self-esteem, and self-confidence. If you put out positive, loving energy when your spirits are high, you will receive it when you need it most., and I speak from personal experience on this point. Try being positive and upbeat to everyone you meet for one week.

The results will amaze you. This one success strategy alone will help you achieve spectacular results!

Look for the Ease

Never forget that work is a measure of results, not effort. Keep this in mind at all times. Got a task to do? Starting a project? What's the easiest way to get it done? If you look for ways to get results smarter instead of harder, you'll find that success really does come easy. Since you haven't expended all of your energy getting some success, you'll have plenty left that you can use to go out and get even more. Imagine getting more for less. You do that while shopping, don't you?

My definition of success is to live your life in a way that causes you to feel a ton of pleasure and very little pain.
Anthony Robbins

Always Expect Success

Go into every meeting, sales presentation, interview, negotiation, training session, etc. expecting nothing less than total success in whatever outcome you're seeking. You may not get what you want, but you will always get what you need. It's important to make the distinction between getting what you need and putting one over on the other guy. The basic idea is to always seek a win-win scenario where all parties walk away feeling satisfied. Just don't get attached to any particular outcome. Learning to see the blessing in anything that may happen is a surefire way to put yourself in a no-lose position.

We tend to live up to our expectations.
Earl Nightingale

You Have the Power

You create your own personal reality. You are responsible for your own life. You can change your beliefs. You can achieve great success with great ease. You can fulfill your life's mission and secure lasting freedom for yourself and your entire family for many generations to come. No one else can do this but you. You've changed your beliefs. Now keep consciously applying these success strategies and you will experience success and freedom that will blow your mind and change your life. Expect success. Plan for it. Then make it happen. You have a great deal of power... and with great power comes great responsibility.

The point of power is always in the present moment.
Louise L. Hay

A Real-World Example

For every disciplined effort there is a multiple reward.
Jim Rohn

My longtime friend Carolyn and I once talked about success and success strategies. During this conversation, she shared her simple method for incorporating every single one of the success strategies I just presented. Talk about serendipity! An enlightened savage is all about achieving maximum results with minimal effort, and Carolyn's method—you be the judge.

Carolyn told me that she used to spend long hours in her law office struggling to get through her daily workload. When confronted by daunting or unpleasant tasks, she spent hours on trivial tasks or even computer games, anything to avoid biting the bullet. The result? She was in her office day in and day out, struggling seven days a week. Here's how she kicked that habit, and how you can, too:

Every morning, list everything you need to accomplish that day. Job, business, and personal errands all qualify, provided they are things you need to do but may or many not want to do. Next, list all the fun enriching things you want to do that day in the separate area provided on the template. Use a fresh sheet every day. The trick is to arrange your tasks in order from your least favorite to your most favorite. In other words, take the task you are most dreading or wanting to avoid and put it right smack at the top of your list. Tasks you enjoy doing go towards the bottom. Again, be absolutely sure to arrange your tasks from least to most favorite.

Having listed everything in order of preference, go do every task in the order you wrote it down. Don't skip ahead unless there's no way to rearrange your schedule. Right off the bat, you've knocked out your least favorite assignment and gotten past it. From there, the rest of every day gets progressively easier and more enjoyable.

Take a closer look at your template. If the task at the top of each day's list is the one you look forward to doing least, then isn't it the one you'd most like to accomplish that day? Here's how powerful Carolyn's system is: Follow this system faithfully, and you will begin each and every day by triumphing over your biggest obstacle.

Imagine how good you will feel, how charged up you will be, and how full of energy and motivation you'll be every day, knowing that you've gotten over the hump and that the rest of your day is all downhill. I bet you'll soon start rearranging your days to knock out your least enjoyable task first all the time.

Use this simple system, and you will come to cherish your least favorite tasks, because they represent freedom for the rest of the day. They represent using obstacles as steps up instead of barriers. You will struggle and procrastinate less. Take it from me, procrastination is a form of struggle.

Here's the best part: You will have more time for the most important person in your life: you. What will you do with that time? Take more vacations? Grow your business? Spend time with your loved ones? Take that class you've been meaning to take? Write a book? It's up to you, because it's your time. How much time? Carolyn estimates she's saving two hours per day. That adds up to twenty days per year! What would you do with an extra three weeks of free time per year? Carolyn embodies what being an enlightened savage is all about: achieving great results with the least amount of effort.

Her income has remained the same. Her clients still receive excellent service that has only improved since Carolyn started taking on each day with a whole new attitude. She has more time to spend with her loved ones. All it took was a very simple adjustment in how she tackled her daily tasks. Again, what would you do with an extra three weeks each year? How much is that time worth to you?

Expect Success!

Memorize all fourteen success strategies and seek ways to use them every day, within the framework I presented in Chapter 15. These tools will help you in your quest to maintain and strengthen your new beliefs. It may feel uncomfortable and artificial at first, like you're trying to be someone you're not. Thanks for the step up!

With an eye made quiet by the power of harmony, and the deep power of joy, we see into the life of things.
William Wordsworth

Questions

Use Caterina's *Success With Ease* strategies, and you will experience a lot more success and a lot more ease in your life. Write down your answers to the following questions:

- **What do you think and feel about the subtle yet profound changes you've made in your mental programming?**
- **Are you committed to doing the follow-up required to nurture and strengthen your new programming so that your old programming won't rise up again?**
- **Will you say, "Thanks for the step up!" whenever you confront an obstacle? How can you be sure of this?**
- **On a scale of 1 (not at all) to 10 (perfectly), how well have you used the success strategy of choosing joy in the past? _____**
- **On a scale of 1 (not at all) to 10 (perfectly), how well have you used the success strategy of living on purpose in the past? _____**
- **On a scale of 1 (not at all) to 10 (perfectly), how well have you used the success strategy of acknowledging others often in the past? _____**
- **On a scale of 1 (not at all) to 10 (perfectly), how well have you used the success strategy of being willing to be uncomfortable in the past? _____**
- **On a scale of 1 (not at all) to 10 (perfectly), how well have you used the success strategy of asking for what you want in the past? _____**
- **On a scale of 1 (not at all) to 10 (perfectly), how well have you used the success strategy of exploring new possibilities in the past? _____**

- On a scale of 1 (not at all) to 10 (perfectly), how well have you used the success strategy of maintaining a positive disposition in the past? _____

- On a scale of 1 (not at all) to 10 (perfectly), how well have you used the success strategy of taking small actions towards your goals in the past? _____

- On a scale of 1 (not at all) to 10 (perfectly), how well have you used the success strategy of openly expressing your gratitude in the past? _____

- On a scale of 1 (not at all) to 10 (perfectly), how well have you used the success strategy of having some fun in the past? _____

- On a scale of 1 (not at all) to 10 (perfectly), how well have you used the success strategy of smiling, laughing, and loving more in the past? _____

- On a scale of 1 (not at all) to 10 (perfectly), how well have you used the success strategy of looking for the ease in the past? _____

- On a scale of 1 (not at all) to 10 (perfectly), how well have you used the success strategy of expecting success in the past? _____

- On a scale of 1 (not at all) to 10 (perfectly), how well have you used the success strategy of owning and using your own power in the past? _____

- How well will you use each of the 14 success strategies from now on? What specific measures will you take to ensure that you do in fact use them?

Chapter 17

Traits of Every Enlightened Savage

Man's main task in life is to give birth to himself, to become what he potentially is. The most important product of his effort is his own personality.
Erich Fromm

This chapter presents the twelve personality traits shared by all enlightened savages. Adopting all twelve traits will make achieving your goals and living your dharma far easier. Find ways to use each of these traits when pursuing any of the fourteen *Success With Ease* strategies from the previous chapter. Do that and you'll find much more success with much more ease.

This list of traits is taken directly from the bestselling *Guerrilla Marketing* series by Jay Conrad Levinson. There is a very simple reason for this. In my opinion, a truly successful guerrilla marketer must, by definition, be an enlightened savage. By contrast, an enlightened savage has all the ingredients s/he needs to become a very effective guerrilla marketer.

As you read this chapter, ask yourself how well you score on each of the twelve traits.

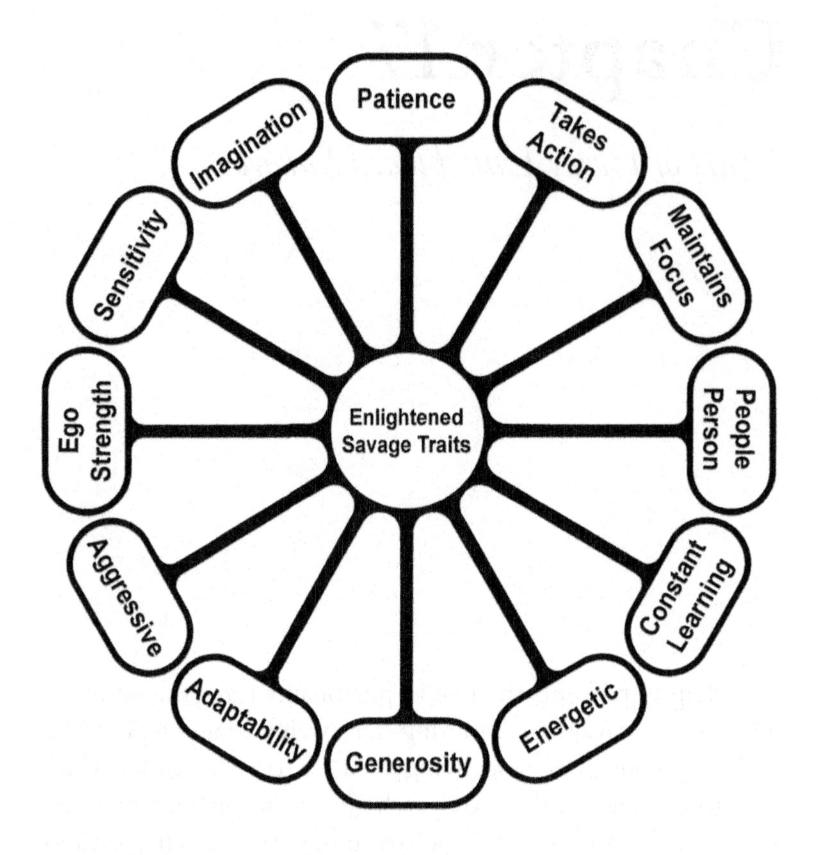

Patience

Patience is the ability to idle your motor when you feel like stripping your gears.
Barbara Johnson

There is an old proverb that says, "Do not push the river; it flows." The river of life has its own flow. Sometimes you'll be swept up in raging rapids that carry you off down unexpected paths. Other times you'll be in calm water and feel like you're not really moving forward. This ebb and flow is part of the great pattern of life. Wanting impatiently for something to happen will only delay that thing from happening. I believe Pastor Douglas Fitch of Glide Memorial Church in San Francisco, California, said it best when he said (and I paraphrase):

> *The Creator is trying to give each of us a miracle every day. The problem is that we place so many demands and conditions on this gift that it either arrives hopelessly twisted out of shape or not at all.*

Patience is the ability to let go of your demands and conditions and accept what comes is. Good things are coming to you if you believe they are. Let them come on their own time, and you will be richly rewarded indeed. Let go of whatever need you have to control things. As we discussed, you have precious little real control over the many things in your life so why bother trying to pretend differently?

Imagination

We are the miracle of force and matter making itself over into imagination and will. Incredible. The Life Force experimenting with forms. You for one. Me for another. The Universe has shouted itself alive. We are one of the shouts.
Ray Bradbury

"What if?"

Enlightened savages ask that question all the time, as do all prey animals. The difference is that you used to ask that question with dread before becoming an enlightened savage. Having upgraded your mental programming, I hope you're asking that question with a sense of eager anticipation. "What if?" is the question asked by an active imagination. Just asking that question forces you to think conceptually about things that may not be now, but which could be in the future.

You have the power to imagine anything you want, positive or negative. The more you imagine positively, the more wonderful things will come your way. It really is that simple.

Sensitivity

Manners are a sensitive awareness of the feelings of others. If you have that awareness, you have good manners, no matter what fork you use.
Emily Post

An enlightened savage is very sensitive to her or his environment and to everything and everyone within that environment. This sensitivity lets you discern subtleties and nuances that might otherwise be missed, and is part of both making conscious choices and your conscious interrupt. Couple this with an active, fertile imagination and you'll be snatching opportunity seemingly from thin air.

Sensitivity is also very important when interacting with your fellow human beings. You will be able to sense and respond to their emotions by reading the many clues everyone gives off in the form of body language, expression, movement, and speech. You will also be able to take others' feelings into account and make them feel heard, understood, and truly listened to at all times—a skill that can have dramatic payoffs.

Ego Strength

Egotism is the art of seeing in yourself what others cannot see.
George V. Higgins

In the chapter on dharma, you learned that your transformation into an enlightened savage might not always find support from people around you. Indeed, you might face severe pressure to slow, halt, or even reverse the changes you've undergone as a result of this program. An enlightened savage retains the strength of ego required to know that they are on the right path and creating the life and dharma they were born to lead. She does this by accepting and loving herself unconditionally, which allows her to let go of her need for acceptance from others.

Your path may require you to counter and/or even distance yourself from unsupportive people. Some of these people may be very close to you by relation or other bond. Remember that you must put yourself first if you are to live your dharma. This is a difficult thing for anyone to do and is why ego strength is such an important part of being an enlightened savage.

Aggressiveness

My passions were all gathered together like fingers that made a fist. Drive is considered aggression today; I knew it then as purpose.
Bette Davis

One of the dictionary definitions of *aggressiveness* is, "the quality of being bold or enterprising." Under this definition, enlightened savages are the epitome of aggressiveness. They know that the Universe won't simply hand them their ideal lives on a silver platter. Living one's dharma requires the boldness to act despite obstacles and naysayers, sometimes to the point of abandoning all that one has in order to obtain all that one needs. As for enterprising, enlightened savages are always looking for the easiest way to achieve their desired results, which often requires creativity and more than a little ingenuity.

This word also has negative connotations that revolve around hostility. Enlightened savages never use hostility to achieve their ends, because they know how critical it is to always treat everyone around them in the manner they themselves wish to be treated. They also never attach themselves to or identify themselves with any particular outcome.

Adaptability

If we don't change, we don't grow. If we don't grow, we aren't really living.
Anatole France

If there is one constant in life, it is change. After all, nothing in this life is ever permanent. The very process of becoming an enlightened savage requires subtle changes to your mental programming that cause profound life changes, like a railroad switch where moving two small pieces of metal causes an entire train to move in a completely different direction.

Enlightened savages know that they are always works in progress, that their process of changing themselves and their realities will continue for the rest of their lives (and probably beyond). As T. Harv Eker says, if you're not growing, you're dying. Embrace the many changes you will experience, look for the oft-hidden blessings in each one, and your life will be blessed indeed.

Generosity

Generosity is giving more than you can, and pride is taking less than you need.
Kahlil Gibran

Look back at your life, and you'll probably see that you would not be where you are today without the generosity of many people who invested their time, energy, and other resources in you. Just as these people invested in you, so must you invest in other people.

Enlightened savages know that giving is the best way to create or enhance the abundance in their lives. They are therefore generous with their own time, energy, and money. This could be as simple as writing free articles containing valuable information that will reward you when readers come to view you as an expert and seek your (paid) counsel. It could be as involved as starting a charitable foundation that gives millions of dollars to worthy causes.

Being generous does not mean that you must fritter away everything you have in gifts and donations. It does mean that you must decide how much you can comfortably give and always give that amount. Over time, this amount will increase, and your generosity should increase to match. Never forget that generosity applied with wisdom and grace is always the best way to increase your own abundance. It really is better to give than to receive.

Energetic

Good luck is the willing handmaid of a upright and energetic character, and conscientious observance of duty.
James Russell Lowell

Enlightened savages are never slouches. They know that living their dharma means pouring every ounce of life energy they have into that pursuit. The beauty of putting this much energy into your life is that it will come back to you several times over. Living your dharma is the gift that keeps on giving, throughout this lifetime—and possibly beyond—in the form of good karma.

Have you ever put energy into something and come away feeling drained? That happened because whatever you put your energy into was not in alignment with your dharma or passions. I'm not saying that pouring energy into your life won't tire you out. Far from it. I am saying that the exhaustion you feel will be a source of joy and satisfaction, which are also forms of energy.

For example, imagine going on a long hike in the woods and returning home completely wiped out. Your physical exhaustion will probably feel wonderful because you are investing in your body's health and reaping the physical, mental, and spiritual rewards for your efforts. Even better, your energy investment will return to you in the form of stronger muscles that will eventually allow you to make the same hike faster and far more easily.

Constant Learning

Learning is not attained by chance. It must be sought for with ardor and attended to with diligence.
Abigail Adams

Constant change and growth require constant adaptation if you are to continue to thrive. Enlightened savages are continually absorbing new information and putting it to good use to increase their abundance and the abundance of those around them.

There are many ways to learn, few of which involve traditional classroom settings. How do you learn best? It could be from reading books, listening to audio programs, watching a show or demonstration, and/or actually performing the task.

People Person

Small kindnesses, small courtesies, small considerations, habitually practiced in our social intercourse, give a greater charm to the character than the display of great talents and accomplishment.
Mary Ann Kelly

No person is an island, and no person can live their dharma without the help of others. Enlightened savages depend on other people for information, guidance, assistance, emotional support, and much more. They actively seek out people who can help them achieve their goals, especially when such partnerships are mutually beneficial.

Maintains Focus

When I chased after money, I never had enough. When I got my life on purpose and focused on giving of myself and everything that arrived into my life, then I was prosperous.
Wayne Dyer

No one can ever be all things to all people. Once they find and accept their dharmas, enlightened savages concentrate on that mission with laser-like focus. They know that digressing means losing focus and diluting their efforts, making achieving their goals more difficult. They know that they will never make everyone happy and thus only seek to satisfy those people who matter, starting with themselves.

Do one thing superbly or lots of things not so superbly. It's a very simple choice. How much of your struggle before becoming an enlightened savage occurred because you lacked focus? That is how much more ease you will have in your life, starting the moment you decide to focus.

Takes Action

Action springs not from thought, but from a readiness for responsibility.
Dietrich Bonhoeffer

All of the programming upgrades and frameworks and strategies in the world are meaningless without action. Enlightened savages are creatures of action because they know that achieving one's goals requires them to actually undergo the required steps. Sitting around wishing you were a pilot will get you nowhere. Attending flight school will.

You can build any personal reality you want (within the limits of the laws of physics), provided that you take all of the steps necessary to bring what you want to life. never attempt to do this through "attraction" or "manifestation." Take action to make your dharma real. Don't attract your ideal life. Realize it!

Questions

Master the twelve personality traits, apply them in everything you do, and you will be far along your path to becoming an enlightened savage. Master only eleven, and you will have a critical piece of the puzzle missing. Master ten, and—you get the picture. Rate yourself from 1 (not at all) to 10 (perfectly) on how well you have integrated the personality traits of enlightened savages into your own life.

- **Patience _____**
- **Imagination _____**
- **Sensitivity _____**
- **Ego Strength _____**
- **Aggressive _____**
- **Adaptability _____**
- **Generosity _____**
- **Energetic _____**
- **Constant Learning _____**
- **People Person _____**
- **Maintains Focus _____**
- **Takes Action _____**
- **How can you improve your mastery of each of the secrets of enlightened savages where you scored less than a perfect 10?**

Chapter 18

The Climb Up

Always continue the climb. It is possible for you to do whatever you choose, if you first get to know who you are and are willing to work with a power that is greater than ourselves to do it.
Oprah Winfrey

Well, you've changed your beliefs and supercharged your expectations. Your life is poised for some potentially major changes. The next step is to get your physical personal reality ready to accommodate these changes. Remember that you purchased *The Enlightened Savage* because you're missing something you want or need in life. In other words, your present situation is not your ideal situation. So how do you begin turning things around?

There's only one way: Make your present situation manageable and take charge of it. The Universe will not give you any more than you can manage. If you want more, you must manage what you do have. That might require sacrifice. If you need to drop some dead wood now to prepare for bigger successes later, then I would argue that this is a sacrifice well worth making. Let's discuss this in more detail.

I drew inspiration for this chapter from a conversation I had with Mr. Andrew Davis, a friend who built a successful network marketing business. He told me that a good number of people are drawn to network marketing out of desperation. The money they have to invest in their businesses is everything they've got left.

These people come to Andrew when they are literally days, even hours, away from losing their homes, cars, furniture, and more. They are drowning in debt and were just scraping by until an injury, illness, layoff, or other event brought their fragile world crashing down around them. They call Andrew thinking that they can buy into his business and earn piles of money overnight. This is not exactly the best position from which to launch a new endeavor. After some discussion, we realized that rich or poor, comfortable or about to lose it all, the process of preparing to experience the success you deserve is the same. Here it is.

Money Doesn't Matter

The person who said money is the root of all evil just flat out didn't have any.
Stuart Wilde

Your financial and material status doesn't matter nearly as much as you think it does. I know this from personal experience. You may be scraping by on a low-paying job, such as Barbara Ehrenreich describes in her book *Nickeled and Dimed.* If that's the case, you're depending on your good health and every cent you can earn just to eke out a meager living. You may be on welfare or dependent on other public benefits. Or, you could be a high-powered executive living in a mansion with every luxury and debt up to your eyeballs. I was a successful consultant, myself. It doesn't matter. Any way you look at it, your house of cards might be one light breeze away from collapsing.

Americans have racked up an unheard-of level of debt; this cycle can only continue for so long before breaking. It's already breaking, as evidenced by the record number of bankruptcies being declared and staggering numbers of foreclosures. Savings are near their lowest levels ever despite rising in the wake of the Great Recession of 2009. Meanwhile, bankers and tycoons are richer than ever, and government seems unwilling to bring them to heel. This is not a pretty picture.

Are you in control of your impulses and your finances, or are they in control of you? The thing about debt is that it gives you temporary freedom at the expense of long-term slavery, because each month you have less and less discretionary income. Whether you're approaching or beyond the point of safe return, or whether you simply want to get started on the path to lasting success, the strategies are the same.

Straightening Your Financial House

Getting your house in order and reducing the confusion gives you more control over your life. Personal organization somehow releases or frees you to operate more effectively.
Larry King

I'm going to give you a fast overview of how to begin getting your finances in order. This brief introduction cannot replace personal consultation with a financial planner. Speaking of which, if you want to learn one of the simplest and most powerful money management systems out there? If so, then you need to attend T. Harv Eker's *Millionaire Mind* weekend. You will learn techniques that will start—and keep—your net worth growing. Thinking of starting your own business or looking for ways to make it grow? Learn from someone who went from making $1.67 per hour in a factory to self-multimillionaire using Jim Britt's *Do This. Get Rich!* programs.

Start Where You Are

You must go after your wish. As soon as you start to pursue a dream, your life wakes up and everything has meaning.
Barbara Sher

The very first thing you need to do is take honest stock of your situation. What's your monthly income before and after taxes? How much is each item you own worth, from your home to your socks? Remember to value what you own in terms of what you could realistically sell it for, not how much you paid for it. Next, how much do you owe in all forms of debt, from charge cards to mortgages? Subtract the debt from your assets. The resulting number is your net worth, or how much you truly own.

Be warned: This may be a negative number! It doesn't matter how much you own if it's all offset by how much you owe on it. If this number is negative, you are in grave danger of serious financial problems up to and including bankruptcy. This realization may come as a shock to you. Or, you may be staring at an eviction notice and knowing all too well the dread that comes with pending ruin.

Cut Back

Dreams do come true, if we only wish hard enough. You can have anything in life if you will sacrifice everything else for it.
Sir James M. Barrie

The next step is to decide if and where sacrifices must be made. But wait: Isn't *The Enlightened Savage* all about success? So then why the heck am I telling you to make sacrifices? Because this is the path to freedom. Yes, you may wind up owning few or no things, but at least you won't have to juggle as much to stay afloat. Remember how we approached your beliefs: We broke through your walls instead of walking around them. The same idea applies here.

Could you hang to everything you have out of some misguided sense that material possessions equal wealth and happiness? Possibly. But since you bought this book, you must already know that your focus on materialism is actually the cause of your present predicament, and that the cause was fueled by your old beliefs! Ditch the goods, and you ditch your problems. You can then begin building back up from a position of relative strength instead of trying to climb with this huge burden on your back. The moment I stopped trying to hang on and let go of my credit cards, car, and house, my life started turning around. I'm climbing back up now—only this time I'm not stressed. Not coincidentally, my net worth is rising faster than ever, and I am taking steps to make sure it stays that way.

How much should you cut back? That depends. If you have enough savings or other available cash to continue your present lifestyle for at least six months without any additional income, and if you are spending no more than a reasonable amount of your current income on debt, then you're all set. You clearly know how to manage what you already have and only need to learn how to build on that with the least possible amount of struggle. On the other hand, if you're about to be kicked out of your apartment with no savings and a few boxes of stuff, then you must take more drastic measures. Let's take the worst-case scenario: You've just lost a low-paying job, and are about to be evicted from your home. Your car barely runs, and there are bigger numbers on your credit card bills than you can bear to look at.

Stay Positive

> *We have been taught to believe that negative equals realistic and positive equals unrealistic.*
> Susan Jeffers

How the heck can you possibly have positive beliefs and expectations while being confronted by homelessness? Well, my friends, if you're in this situation, then at least you can't fall much lower. This may seem like small comfort until you realize that you've got no place to go but up. And you can, and will, go up, if you remain true to your new beliefs and expectations and constantly practice the success strategies we discussed in the previous chapter. Chapter 9 gave you some insight into what you should be doing in order to discover your mission in this life. That means that your path is before you, and that you are already ahead of many people. If your situation isn't this dire, then count your blessings!

Even more importantly, your situation is neither good nor bad because nothing in your life is ever good or bad by itself. You choose to see it as good or bad. Choose to see the blessings, lessons, and opportunity in your life and you will find yourself less stressed and more clear-headed than you may be able to imagine right now. Let go of the need to control what happens and the need to be accepted by the mythical Joneses, and you'll instantly lose a lot of the heavy baggage that's been weighing you down.

Budget, Budget, Budget

Some couples go over their budgets very carefully every month. Others just go over them.
Sally Poplin

Create a budget for yourself. Add up everything you spend money on during a typical month. Include absolutely everything. Leave nothing out. Now that you have your monthly outlay in front of you, examine each item and ask yourself: Is this contributing to my success or to my stress?

If the item in question is contributing to your mission and helping you achieve the goals you laid out at the beginning of this seminar, by all means keep it. If not? Dump it. Anything that isn't helping you grow is helping you die, period.

The fact that you're keeping critical things, such as power and phone, doesn't mean that you can't cut back. My phone bill was over $200 every month. I switched to a service with unlimited long-distance calling, and instantly knocked over $125 off my budget. Switching lights off and adjusting your thermostat just a few degrees can yield tremendous savings on your power bill. Do you really use all your cellular minutes? It may be cheaper to get a smaller plan and pay the occasional overage.

Budgets are for cutting, that's why you set them.
Dr. Laurence Buckman

Do you really need to drive everywhere? How about walking or bicycling to your errands? Your health and energy levels will jump, and you could save many dollars per month on gasoline, oil, and maintenance. Speaking of your car, is it truly essential? If so, are all the fancy gadgets in it really necessary? You might be able to save lots of money by trading in that sports car or SUV for a more modest vehicle. Do you need to eat out as often as you do? Yes, you must nurture yourself and prime your ability to receive, but if your dining habits are threatening your survival, then you should consider less costly (or even free) ways to enjoy yourself. My friend Shel Horowitz's book, *The Penny Pinching Hedonist*, is packed full of great ways to enjoy and nurture yourself without spending an arm and a leg. We'll talk more about this later. Meantime, let's continue slashing and burning your budget.

Beware of little expenses. A small leak will sink a great ship.
Benjamin Franklin

As you can see, there are many ways to trim your expenses without sacrificing your true quality of life. OK, your new car might not be as cool as your previous ride. What of it, if that minor sacrifice gives you more freedom to enjoy what really matters in life? I sold my car and never looked back. When I need a car, I use my handy car-sharing membership.

Housing: If you have one or more mortgages, can you refinance them? If yes, will doing so save you money? Be absolutely sure to figure in things like closing costs and prepayment penalties! If you're renting, can you find an acceptable home for less money? You might have to give up a spare bedroom, put the kids in bunk beds, or forego the great view and golf course. But hey, at least you've got a roof over your head.

Now we come to debt. If you own a home, you might be tempted to pay off your credit cards, cars, vacations, etc. by taking out an equity loan. This may save you money each month, but the long-term costs will almost always far outweigh the small savings. That vacation you took? As wonderful as it was, do you really want to end up paying for it four times over for the next thirty years? Besides, your mortgage payment will increase, making it harder to keep your home in the event of an emergency. Do you really want the that cruise to that tropical paradise to end up forcing you from your home? Also, if you're underwater on your mortgage, then walking away from your home might be the best thing to do. Big corporations don't hesitate to walk away from their problems, and neither should you. Walking away from my home was an extremely liberating experience.

Do not accustom yourself to consider debt only as an inconvenience. You will find it a calamity.
Samuel Johnson

If you have credit card debt, there are a number of non-profit services that can negotiate a lower interest rate to reduce your payments and pay off your balance far sooner than ordinarily possible. Those can be a very attractive option. Car payment too high? If you own multiple vehicles, consider selling one or more of them. If you own one car, consider trading it on a less expensive model.

The key here is to be as ruthless as possible given your situation. If you have a healthy savings account and little debt, you might choose to make minor cuts. Faced with imminent homelessness? Slash and burn.

Income

If one wants to get out and stay out of debt he should act his wage.
Anonymous

OK, enough about your expenses. Let's look at your income. How much do you make? Is it enough to cover your reduced expenses? If not (or if you have no income coming in), how can you fix that situation?

First, are there any jobs available that pay more than what you earn? If you're self-employed, there are many books, seminars, and training programs to help you increase your revenue and, even more importantly, boost your bottom line. Jim Britt's *Do This. Get Rich!* has helped many people achieve and surpass their financial dreams (including me).

Get Help

Be bold and mighty forces will come to your aid.
Basil King

If needed, do you qualify for any type of assistance such as unemployment, disability, workmen's compensation, welfare, food stamps, state medical coverage, public housing, or charity? No one likes to use these services, but they exist for just such emergencies, and this is no time for pride. Can friends and family pitch in? I know we're not talking about an ideal situation, but the whole point of this exercise is to remove as much pressure from your life as possible.

This may not be easy. No one wants to call a creditor to seek alternate payment options. No one wants to go the local food bank for help. Selling your new SUV is a hard thing to do. Whatever your situation, sacrifice is never easy. And that's just the point: What separates people who get what they truly want from life from those who don't? Those who rise to the top act in spite of their fears, doubts, and insecurities, while those who sink surrender to them. You've surrendered long enough. Forging ahead may require you to eat some crow, but you will emerge far better for the experience. Put it this way: Eating sure beats starving!

I do need to stress, however, that the financial considerations presented here are intended only as discussion points. Every situation is a bit different; it's impossible for me to give you specific advice for every possible situation. Always consult a qualified financial advisor before making any decisions.

Again, I highly recommend T. Harv Eker's *Millionaire Mind Intensive*, a 3-day seminar that provides one of the world's simplest, most effective money management strategies and helps you get rid of any negative core beliefs you may have around money. I personally attended this event. Believe me when I tell you that it is life-changing. This is a free event. If you're serious about turning your financial situation around, then you literally can't afford not to go!

Examine Your Relationships

The quality of your life is the quality of your relationships.
Anthony Robbins

If you are involved in any relationships that challenge or otherwise don't mesh with your new beliefs and expectations, then now is the time to reevaluate them. List each person with whom you have a significant relationship, and think about each one. You'll probably find that you can fit them into one of several categories.

If your relationship with a person affirms your new beliefs and expectations, great! You have a true friend and soul mate. If not, is it because this person is someone who has not yet experienced the programming-changing processes we covered in Chapters 11 through 14? Or is there some history or unfinished business between you?

If the former, please recommend *The Enlightened Savage* to this person to help them get started on their own paths to success. If the latter, then I recommend that you conclude your business. If your relationship can evolve to a mutually constructive point, then by all means work towards that end. If not, then part company. This may be hard if it's a spouse or a family member, but it may still be the healthiest option for you and for them. Remember that the best path forward is not always the easiest. You may even find that people actively resist your changed beliefs. All I can tell you is that people always act for themselves, never against anyone else. Any objections you receive to any change of profession, residence, business, etc.

only reflect that person's perceived needs, and have nothing to do with you. As always, the one question facing you is: Surrender or succeed? It's up to you.

Keep Climbing

Don't wait for someone to take you under their wing. Find a good wing and climb up underneath it.
Frank C. Bucaro

Well, here we are: You've changed your beliefs and expectations and have removed the dead wood from your life. I focused primarily on the financial aspects because money tends to be behind most stress and strife in life. Getting rid of that stress and strife gets you on your way to true freedom. I also touched on your relationships. This lifetime is just too short to stick with those who do not support you. It's bad for you and bad for them, a true lose-lose proposition. I don't want you to ditch all your friends. Just be prepared to make sacrifices in the name of achieving your life's goals and dharma. You're all you have. Without you there is nothing.

Questions

Sacrifice is difficult, no doubt about it. Remember to focus on the desired outcome: Free yourself from everything and everyone holding you back. You've had enough of that in your life and deserve a change for the better. Changing beliefs and taping up a card with success strategies on it is easy, but sacrifice is what tests those beliefs and strategies. The extent to which you can make any necessary sacrifices and move towards your life's goals and mission is the extent to which you'll succeed. If your load is too heavy, lightening it will help your life change for the better. Manage what you have and you will get more. It's that simple.

Write down your answers to the following questions:

- **Is your current living situation stable? If not, why not? Be as specific and detailed as possible.**

- **Do you feel that you are in control of the situation? Why or why not?**

- **How can you take control of the situation? What resources do you have that can assist you?**

- **Create a detailed monthly budget. Is it sustainable? What expenses can you reduce, if necessary?**

- **Are you facing an imminent calamity such as eviction, bankruptcy, medical bills, or personal debt? What specific steps can you take to mitigate the situation?**

- **Do you realize and accept the fact that you created your current personal reality using your old programming and that you have the power to create almost any personal reality you want?**

- **How can you maintain your progress towards becoming an enlightened savage in the face of any challenges you are currently facing/ Be as specific as possible.**

- How much of your current situation has been caused by your not accepting or living your dharma?

- How is your current situation a clue to the life you should be living?

- Do you own the fact that your current situation represents your 100% successful execution of your old programming?

- Do you understand that the fact you are in this situation means you can be in any other situation you want?

- What specific commitments can you make to begin to change your situation?

- Who can you turn to for help?

- How can you reduce or eliminate any debts you might have?

- How can you boost your savings and investments towards creating financial freedom for yourself?

- What can you do to boost your income?

- Are you surrounded by people who support you and your path to living your dharma? If not, how can you find and meet kindred spirits to help you?

- Are you willing to distance yourself from everyone who is not supportive of your life's path? If not, why not?

- On a scale of 1 (not at all) to 10 (perfectly), how would you rate your ability to manage what you have? How can you improve your skills so as to open the door to receiving more?

- How committed are you to creating the life you want instead of the life you have?

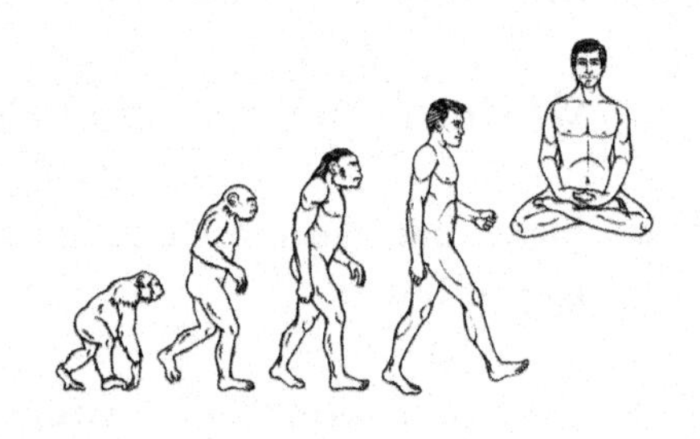

Chapter 19

The Next Level

To have a quiet mind is to possess one's mind wholly; to have a calm spirit is to possess one's self.
Hamilton Mabie

By now, you may be wondering what the heck to do next. Remove all the noise, clutter, and obstacles from your life and the result can be awfully quiet—too quiet. If you've had to make major changes or sacrifices in your life, you may be questioning your resolve, and/or wondering if you're doing the right thing. You may also be wondering what to do with all the time you suddenly have on your hands. You've bitten the bullet and done the hard work. So what's next?

Consider the experience you've just been through as demolishing an old building to make way for a shiny new edifice. You're now staring at a figurative vacant lot, and your new task is to build another building. In other words, it's time to move towards achieving your goals. Earlier chapters in this book helped you define your internal, external, and commercial goals and gave you a few ideas on how to start tuning into your dharma. Your old barriers to success are gone. The future you are meant to have is yours for the taking. Go for it! But how?

Your Goals Matrix

You've already drawn a picture of where you want to be, outlined your important life's goals, and cleared away the obstacles by changing your negative core programming. Your next step is to list your goals in order of priority from most to least important, being sure that you either have or can get the resources you need to accomplish that goal. For example, if you want to become certified to fly charter planes, then you'll need to budget between $10,000 and $15,000 and hundreds of hours. This means that you must be earning enough money to cover both your living expenses and your flight training. Suddenly, your goal of flight becomes linked to a financial goal. I trust you see why cutting off the deadwood makes your job so much easier.

One should act in consonance with the way of heaven and earth, which is enduring and eternal. The superior man perseveres long in his course, adapts to the times, but remains firm in his direction and correct in his goals.
I Ching

If your goal is to write a book, then you'll need to find time to focus on writing. This will be time that you can't invest anywhere else, time that keeps you from achieving your other goals, unless those goals are linked to your writing. Each goal has dependencies and links: A dependency occurs when one goal must rely on another. For example, the goal of becoming a commercial pilot may depend on achieving your goal of earning more money. Working toward this goal might help your fear of heights or trying new things. Each goal you strive for both helps and hinders your other goals. This matrix of dependencies and links might even be circular. For example, your cousin the tour operator might have promised you a great job as a pilot, once you get your license.

Go online to www.dawnstarbooks.com/tes/materials to obtain a blank Goals Matrix ready for you to fill out.

In this case, you must invest the money to pay for your training, but may be able to recoup that and more as a pilot, once you reach that point. We'll talk more about links and dependencies in a few minutes. First, however, let's talk about milestones.

Marking Your Journey

As the name implies, a milestone indicates another step along a journey. The milestones on the way to your goals are the steps you must take in pursuit of those goals. Let's use our pilot example: The first step is to head down to your local airport and select a flight instructor for an introductory flight. You must then obtain a flight physical in order to obtain your

We advance on our journey only when we face our goal, when we are confident and believe we are going to win out.
Orison Swett Marden

student license, which you'll need in order to fly solo. After that come ground school, flights with your instructor, solo flights, and eventually your first major milestone: your private pilot license. From there, you'll want an instrument rating before finally getting your commercial license. If your cousin flies twin-engine aircraft, you'll also need a multi-engine rating.

Writing a book? Research your topic until you can live and breathe it, then create a synopsis. From there, create your outline, which is the book's skeleton. You can then put the meat—the actual writing—on those bones. Meanwhile, you're learning about the publishing industry and deciding how best to publish and market your product once it's complete.

The same basic process applies to all goals: Define milestones and take them one at a time. These milestones can involve knowledge, time, money, just about anything. Each milestone has its own links and dependencies.

Mapping Your Goals

Somewhere there is a map of how it can be done.
Ben Stein

If you mapped out each of your goals and drew all of the links and dependencies between all of the milestones, you'd end up with a page of scribbles—and that would only discourage you. So how should you go about doing this?

Start with your most important goal. None of us knows how much time we have on this Earth, so you'd best get to the things that matter right quick. The *Questions* section at the end of this chapter has space for you to list your goals with your most important goal on the first page.

Next, list the major milestones on the way to each goal. If we're talking about flight training, those milestones might be student license, private rating, and instrument rating. List these milestones in order. Don't know what the milestones are? Research them. There is no harm or shame in not knowing how to get where you're going. The only shame is in not finding out. Remember your success strategies. Following them means asking lots of questions. If your interest and passion are for real, then most of the people you ask will be absolutely delighted to help you. The energy you radiate comes back to you. If someone you approach is less than receptive, just remember that the rejection has nothing to do with you.

Researching major milestones might reveal that they have their own milestones, which I call minor milestones. The first time you solo an aircraft is a major milestone on the way to getting your private pilot license, and a minor milestone on the way to your commercial ticket. While researching, be sure to inquire about each stage of the process so as to capture every major and minor milestone.

You. too, can determine what you want. You can decide on your major objectives, targets, aims, and destination.
W. Clement Stone

Having mapped your goal, find the dependencies and links for each major and minor milestone. Generally, the dependency in achieving a major milestone lies in achieving its associated minor milestones, while the links go to the next milestone. For now, confine your listing to those beliefs that are specific to that goal. For example, concentrate on the flying without bothering about the cost. Your workbook lists these as internal links and dependencies. Focus here for now.

Repeat this process for each of your stated goals. When you're done, you'll have a guide that shows you how to accomplish the things that are most important to you. Your research may take hours, days, weeks, or more to accomplish. That's OK, because you're investing in yourself and your future. You want to get places and do things and are creating the map to take you there.

This map allows you to go the maximum distance with the least amount of effort—in other words, to move your dirt pile with a tractor instead of a kiddie bucket and shovel!

Take Action

Action cures fear, inaction creates terror.
Doug Horton

Committing a goal to paper and committing it to action are two different things. There's also one additional catch: You may need to work on multiple goals at once. If you're writing a book, then you probably need to be doing something else for income. If you're learning to fly a plane, then you must earn the cost of your training in addition to scheduling your flight time. If you don't have enough income, you'd better start saving and/or making more money. Your goals might not be financial. That is perfectly fine, as long as you remain focused on both your goals and the means to achieve them.

Stay Organized

Don't agonize. Organize.
Florynce R. Kennedy

Air traffic controllers have a tough job. They must fill slots in the sky with airplanes moving in different directions at different altitudes and speeds. This is particularly true in metropolitan areas such as San Francisco, which has three major airports, a military field or two, and several smaller airports. Put two or more airplanes into the same slot, and you have the recipe for disaster. Fill too few slots, and delays literally ripple across the nation and possibly beyond. The controller must therefore put one airplane into each available slot. Not zero. Not two. One. If traffic exceeds the number of available slots, they must slow the flow. Should there ever be more available slots than airplanes, everyone can breathe a little easier.

Life is kind of like the sky near a cluster of busy airports. You have big things happening quickly, others more slowly, something else is happening that you need to attend to, and everything is coming from different directions and going to different places. You must make sense of this mess. If you live in a metropolitan area, head over to the airport some evening and watch the planes landing and taking off. You'll see several rivers of metal merging into a neat, evenly-spaced line of planes landing one right after the other. You'll see another river of metal lifting off and splitting into streams as planes fan out in different directions. This is the result of teams of people in the air and on the ground working together to orchestrate a smooth flow. One mistake, and the system turns into chaos that controllers euphemistically term an "air show," with planes scrambling all over the sky to stay out of each other's way. Same with your life: Either get ahead of your goals and obligations and spend a little time working them into a neat orderly flow, or spend all your time forever cleaning up the mess—it's up to you.

Don't mistake activity for achievement.
John Wooden

First, you need to know how long it will take to achieve your goal and how long you're willing to allow yourself. If your goal involves learning, how fast should you go to maximize your knowledge without being rushed, or going so slowly that you waste time relearning things you've forgotten? What external factors contribute to this? For each goal, set a completion date.

If a goal is ongoing, set deadlines for each milestone, and keep setting new milestones as you go forward. For example, learning to play the piano can take a lifetime. I used to be a pretty decent player. Years of neglect later, and today I can barely read sheet music. If you're not growing, you're dying. Set completion dates for each goal, then set reasonable dates for each major milestone.

Resist the temptation to set dates as absolutes, such as "I will obtain my private pilot's license by July 15th, 2005." Stick to relative dates, such as "I will obtain my private pilot's license within six months of starting flying lessons." Don't know how long something takes? Ask! Never undertake anything without a clear idea of the investment and payoff involved. You probably can't achieve all of your goals at once. This is why you must have priorities, and why I recommend that you set relative dates. There's another reason too:

Achieving Multiple Goals

The more specific and measurable your goal, the more quickly you will be able to identify, locate, create, and implement the use of the necessary resources for its achievement.
Charles J. Givens

So far, we've looked at individual goals as isolated entities with no external links or dependencies. Ah, if only it worked that way; but it doesn't. Humans are holistic beings; what affects one area of our life affects them all. The next step is to connect our goals into a matrix of links and dependencies. Again, links help you achieve your goals, while dependencies are what you must accomplish before achieving your goals. Your workbook contains a sample goal matrix. It may sound complex, but it's much easier than it might sound.

Start connecting goals to each other by drawing color-coded lines. Look at each goal and decide if it can help you achieve other goals or if it depends on something else. Don't look at internal links and dependencies. The point of this exercise is to find the most efficient way for you to achieve your goals. This may seem confusing, but please bear with me. You'll soon see where this is going. Ready?

For each goal, draw a colored arrow (such as blue) to connect it to every other helped goal. For example, flying a plane solo might signify achieving your goal of getting past your fear of heights. In other words, the flight milestone is also the successful end of your fear goal. Traveling might fulfill a research milestone for your novel. The possibilities are endless. After doing that, use a different color (such as red) to connect goals to their dependencies. In this example, the final milestone in eradicating your fear of heights might just be to take off in an airplane alone.

As you do this, you might discover a few things: If Goal A depends on Goal B, then clearly B helps A. Some goals will be dependent on others and helpful of still others. Maybe travel depends on curing your fear of flying and helps your book writing. You may also find that some goals depend on yet other goals and milestones that have nothing to do with your presently listed goals. For example, you may need to add income as a goal. If vacations are extremely important to you, then the income becomes important, not for its own sake, but to facilitate your primary goals.

The Goals Matrix At Work

Make no little plans. They have no magic to stir men's blood and probably themselves will not be realized. Make big plans. Aim high in hope and work. Remembering that a noble, logical diagram once recorded will not die.
Daniel H. Burnham

All set? Take a look at your goal matrix. You're looking for a bunch of blue arrows leading away from one of your goals. This is probably the one that best helps you accomplish your most important goals while being easily attainable. For example, if Goal A has three blue arrows leading to Goal B, while Goal C has six arrows, then Goal C is what you want to focus on. Circle the goals with the most arrows leading away from them.

Next, find the goal with the fewest red arrows pointing away from it. This is the goal with the least number of dependencies, meaning it poses the fewest challenges. Circle this one as well. Next, which goal has the most easily resolved dependencies? Buying a house might have one dependency, increasing income, while writing a book only requires concentrated blocks of time for research and writing. Circle the goal with the easiest dependencies.

Time for the judgment call. If you circled the same goal a few times, your path is clear. What will probably happen is that you'll wind up circling two or more goals. Where you go from here is completely up to you. If writing a book is more important to you than buying a house, write the book. If getting over your fear of heights is more important than income for flying lessons, then find a different way to purge that fear. Let your goals and their priorities be your guide. You have what you want to do laid out in front of you and can see exactly how everything fits together.

Want to knock some smaller things out of the way so you can focus on bigger and better things? Want to learn perseverance and concentration on the easy things before moving up? Fine. Want to dive right in and move towards your major goals as soon as possible? Go for it!

When Answers Don't Come

Your path may not be clear immediately, nor will it always remain clear as you move forward. No problem; keep affirming your beliefs and expectations using your success strategies, and keep listening for your call. The answer will come if you seek it. Just be sure to seek the answer on its own terms, being open to what comes and always looking for guidance and help. Hey, does this sound an awful lot like your success strategies, or is it just me? If you seek the answer and open yourself to whatever comes, you will get your reply. It may not be the one you want, but it will always be the one you need.

He who knows all the answers has not yet been asked all the questions.
Anonymous

Scheduling

Scheduling is the final piece in this puzzle. Buy a day planner. The Appendix in this book includes one for your first 90 days following the Enlightened Savage program. After that, you're on your own, because I want buying a date planner to be an integral step towards accomplishing all of your goals. A day planner lets you see your goals and track appointments, milestones, expenses, mileage, and much more. You may also buy a desk or wall calendar if you wish, or use one of the many calendar programs on your computer or PDA if you're so inclined. Just make absolutely sure to have an effective calendar in place and by your side at all times.

There can't be a crisis next week. My schedule is already full.
Henry Kissinger

OK, you've bought the day planner. What should you do with your new toy? Well, if you know that obtaining a pilot's license takes about six months, and know when you'll have the money to do it, flip your book to the correct page and write "Start Flight Lessons" in big letters. Then speak with your instructor about suitable milestone dates. You may be able to work on some, or many, goals at once. If so, your matrix will help you visualize when to start each goal and when to finish milestones that will have a ripple effect across other goals.

Don't Strain Yourself

Some people develop eye strain looking for trouble.
Anonymous

Be very careful not to overextend yourself. There is a fine line between dawdling and active productivity, and an even finer line between getting the most results with the least effort and struggling. Everyone has a sweet spot that balances effort and leisure for maximum effect. Find yours and resolve not to over- or under-commit yourself.

Questions

Please visit www.dawnstarbooks.com/tes/materials to download a copy of the Goals Matrix and take all the time you need to map out your goals and their dependencies and links. Create your matrix, plan of attack, and schedule. Then get to it! The sooner you begin, the sooner you'll see progress, and the sooner your life will begin taking huge steps for the better. Write down your answer to the following question:

- **How well did the Goals Matrix help you plan your path forward?**

Chapter 20

Charge!

When placed in command, take charge.
Norman Schwarzkopf

You've mapped out your goals and formed your action plan. Now it's time to implement everything you have learned, by achieving those goals and living the life you want and deserve to live. If you're familiar with Jay Conrad Levinson's *Guerrilla Marketing* series of books, this chapter will sound very familiar—as it should, because these steps are identical to those used when executing a guerrilla marketing plan in a business. The ten-step accomplishment process in this chapter will help you accomplish any task in your personal and/or business life.

The following diagram lists all ten steps of the accomplishment process. Think of it as pouring and setting a solid foundation before putting up the walls, roof, and finishing details. You may find yourself wanting to forget the preambles and get on with the task or goal at hand. Here is where a little patience and investment will pay huge dividends, but don't get caught in analysis paralysis. Don't rush off half-cocked, either. Above all, don't let yourself get attached to obtaining any one outcome. Let go of whatever need you may have for control.

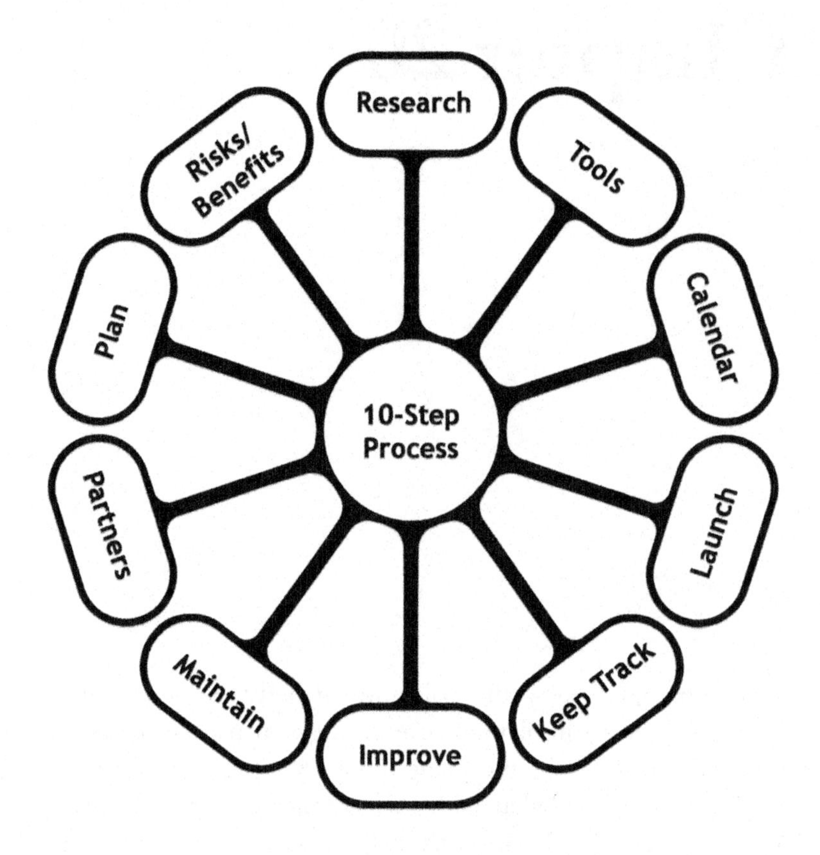

Research

Knowledge is power. The more information you have, the stronger your position. Never commence any major undertaking without a thorough understanding of the steps involved, because this understanding is what will help you create the easy weekly milestones and rewards to keep you on track towards achieving the goal. You will also want to know exactly how much each step of this goal will cost, and how it fits into your overall Goals Matrix, as described in Chapter 19.

Research serves to make building stones out of stumbling blocks.
Arthur D. Little

The key in this all-important first step is to obtain enough information to make a truly informed decision without becoming stuck in "analysis paralysis," where you are so caught up in your research that you never take the next step in the process. Never rush into anything without having all the information you need at hand.

Risks and Benefits

We don't grow unless we take risks.
James E. Burke

What exact benefit(s) will come from accomplishing this goal? The second step of the accomplishment process is where you take the time to list every benefit you can think of. If you like, you can even do this on a per-milestone level. Make sure not to confuse benefits with features. For example, accomplishing one goal might mean that you get a new computer. This is a feature. The specific thing(s) you will be able to do with this new computer that you can't do now are the benefits you should put on your list.

What risk(s), if any, come with accomplishing or attempting this goal? If you're taking up skydiving, one risk might be that your parachute could fail to open, with drastic consequences. If you're getting a new laptop computer to store your company information on, one risk might be that it could get lost, stolen, or otherwise compromised, thus risking potentially sensitive data.

Place the benefits and risks in separate columns on the same sheet of paper, and then examine them side-by-side. Decide the approximate probability that each item (benefit or risk) on your list will occur. Once you have the numbers in front of you, you'll be able to make an informed decision about whether or not you should proceed.

This is a tricky question, because on one hand you created your own personal reality and can create just about any outcome you like (within the laws of physics, of course). On the other hand, it may be best to set some goals aside for later so that you can practice on easier ones first. You must also decide how each potential goal or task fits with your life's mission or dharma. If you absolutely must learn to skydive, then the risk/benefit equation is different than if learning to skydive is just a passing hobby.

Gathering your information and performing this basic risk analysis gives you the tools you need to make a go/no go decision. If you decide to proceed, you will have a solid foundation for moving forward. If not, you will have invested a minimum of time and effort on the matter. Either way, you have set yourself up for the easiest possible outcome, and invested far less than you might by simply floundering.

Here's a hint: The links and dependencies in your Goals Matrix are a great place to begin looking for risks and benefits.

Tools

For the third step of the accomplishment process, what tools do you have at your disposal for accomplishing this task or goal? What additional tools do you need, and how can you obtain them? Here again, your Goals Matrix may give you valuable insight into what you have and what you need.

One of the greatest and simplest tools for learning more and growing is doing more.
John Roger

When selecting the tool(s) you'll use, do your utmost to select only those tools that will get you the most results (work) for the least effort and cost. Ideally, these tools should last you a long time. For example, if you are marketing a new product, a TV commercial might be very easy; however, the money you invest in that one spot might be better invested in direct mail or social media that reaches prospective customers several times over a long period.

Always think simplicity, ease, and longevity when selecting your tools. You'll accomplish a lot more with a lot less.

Plan

The fourth step of the accomplishment process is where you map out how you plan to accomplish your goal or task. Create a time line for accomplishing each step, and list the tool(s) you'll need for each listed step. If you need to acquire any tools such as material, certifications, books, instruction, etc., then be sure to include dates by which you must order or otherwise obtain these items.

It's not the plan that is important, it's the planning.
Graeme Edwards

Your completed plan should resemble the steps outlined in your Goals Matrix, except that it will be much more detailed. Ideally, you'll want to create a map of the steps you'll take each week, so that you can get into the habit of taking one easy step at a time and rewarding yourself for every bit of progress. The process of laying out your plan will also help you to focus only on the step you're currently taking, without getting distracted by the steps you've already taken or have yet to take.

Calendar

I've been on a calendar, but I've never been on time.
Marilyn Monroe

You'll want to create a calendar to track your efforts and their results for the fifth step of the accomplishment process. Lay this calendar out with 12 rows, one for each month, and the following five columns:

- **Month:** List the calendar month.
- **Steps:** List the step(s) you will take during the listed month.
- **Tools:** List the tools you'll use to accomplish those steps.
- **Cost:** List the cost, in time and money, of accomplishing the steps for that month.
- **Grade:** When the month is over, give your efforts for that month a letter grade from A (excellent) to F (no progress).

If you are using your calendar to achieve a specific goal (such as obtaining your private pilot license), you will be able to see instantly how your efforts are proceeding. If all is going well, you'll have the proof in front of you that your new enlightened savage programming is working exactly as intended. If not, you'll know that you need to take a good long look at what is going wrong, so that you can get back on track as quickly as possible. This is not about beating yourself up or finding fault or failure. Rather, it's about helping you identify areas when you may need to pay extra attention, or where you need to take another look at your mental programming or practice letting go and using your conscious interrupt. You may even determine that the goal or task at hand isn't contributing to your dharma and decide to drop it. In any case, this calendar gives you a quick and easy self-diagnostic tool that you can use at any time.

Being rich is having money; being wealthy is having time.
Stephen Swid

What if you've created a calendar for repetitive tasks, such as marketing your business? No problem. Keeping track of what you're doing, and how, and the results you're getting will help you fine-tune your process into a well-oiled machine. This may take time; in fact, many people report needing two or three years—or even longer—to get their calendars for their recurring tasks to show straight A+ grades.

Partners

> *When a match has equal partners then I fear not.*
> Aeschylus

Remember that no person is an island. Each of us depends on everyone else to varying degrees. This is why the sixth step of the accomplishment process is finding out who you can turn to for help. This list can include friends, family, coworkers, vendors, other businesses, coaches, consultants, and much more. Identify the people who are best able to help you accomplish this goal or task with less effort and much more ease.

Launch

> *You don't make progress by standing on the sidelines, whimpering and complaining. You make progress by implementing ideas.*
> Shirley Chisolm

The number seven is widely considered a lucky number. It is no coincidence that step seven of the accomplishment progress is to launch your effort. All of your efforts so far have been directed towards this exhilarating moment.

If you are setting out to accomplish one of your life's major goals, keep in mind that you may well find yourself challenging the boundaries of your old comfort zone. This is why you've broken this goal down into easy weekly steps that will allow you to move at a comfortable pace and enjoy the journey without worrying about the goal. Think of it as inviting the gopher out of his burrow at high noon one baby step at a time. What will keep the gopher coming out when his prey instincts are telling him that hungry hawks are circling overhead? Read on...

Maintain

> *Another flaw in the human character is that everybody wants to build and nobody wants to do maintenance.*
> Kurt Vonnegut Jr.

The eighth step in the accomplishment process is to maintain your efforts. You will encounter obstacles. All of them will be of your own making, because you alone create your personal reality. The question is how you will handle each of these obstacles. Will you say "Thanks for the step up!" or will you let your challenges defeat you? Remember, your challenges are only as good or as bad as you allow them to be.

There is a very simple way to help keep your progress on track. The first thing is to break your task or goal into simple steps that you can accomplish each week. Doing this is the metaphorical equivalent of easing the gopher out of the burrow one baby step at a time. This gradual process lets the gopher see that there are no hawks overhead in a manner that lets him scoot back underground at any time—a very different thing than trying to yank the gopher into daylight all at once, which would be a very traumatic experience indeed.

The most important thing you can do to keep yourself moving forward is to reward yourself for each step you take. You must reward yourself with every step, because this is how you will program yourself to associate joy with making progress towards living your dharma. The rewards you give yourself need not be expensive or extravagant. They must, however, be both original and meaningful. For example, if you frequently go out to dinner, then don't reward yourself with yet another meal out. Your reward could be as simple as watching a sunset or reading a book.

What if you don't accomplish your step for the week? Don't punish yourself. You've had enough punishment in your life; you certainly don't need any more. The only consequence to not completing your step is that you won't get the reward.

If you are working on multiple tasks/goals in the same week? Give yourself one reward for every step of everything you accomplish. If you're making progress on six projects, then give yourself six rewards, one for each project.

Keep Track

Success has always been easy to measure. It is the distance between one's origins and one's final achievement.
Michael Korda

The ninth step of the accomplishment process is to measure your progress using your calendar. If you are accomplishing your steps each week and enjoying your rewards, you can give yourself straight A grades.

Many of my clients make solid progress for a while, then start having difficulty keeping going. I believe this occurs because they've reached a point where the fear of leaving their comfort zone has risen to match the joy of accomplishment they feel. This is perfectly natural, since it's caused by remnants of your old programming realizing that you've reached your former limits. Chances are very good that you will experience solid progress for a while before hitting a wall and stalling.

You may be very tempted to abandon your task or goal and accept defeat. Thanks for the step up! As an enlightened savage, you know that the path to greatest growth is the path with the most obstacles. How can you proceed despite your mounting fears and insecurities? Easy: Take smaller steps. All you need to do is break your planned steps down into smaller, easier chunks. Yes, this will affect your completion date, but that's not important. Slowing down and making sure to reward yourself equally for the smaller steps will ease you past your fears.

This slowdown seems almost inevitable, so don't feel bad if it happens to you. The good news is that you will get past your barriers if you slow down and remember to reward yourself. Even better, overcoming your obstacles will be one of the most exhilarating things you'll ever do. As your fears subside, you'll feel inspired to move faster and faster. No problem: Simply set bigger steps for yourself. In most cases you'll find that the acceleration you experience as you enter the home stretch will more than make up for any delays you experienced when you confronted your fears.

Improve

As you proceed each week, keep asking yourself how you can improve both your progress and your process. This is where you calendar comes in very handy, especially if you are tracking a repetitive or long-term task. For ongoing efforts, it could take two or even three years (or more!) to start seeing straight A grades on your calendar. The key is to identify what's working and what isn't, so that you can eliminate the things that aren't working without affecting those that are.

Practice without improvement is meaningless.
Chuck Knox

Questions

Apply the ten steps outlined in this chapter to everything you do, and you'll amaze yourself with how much you'll accomplish and how easily you'll do it. The questions below are the template you should use when setting out to accomplish a goal or milestone. Write down all of the following information in detail for each milestone you want to accomplish:

- **Research: What sources of information exist to help you accomplish this milestone? What information do these sources contain? How can you use this information to your advantage?**

- **Risks and Benefits: What are the risks you will face when moving forward? What benefits will you obtain by completing this milestone?**

- **Tools: What tools (people, equipment, money, etc.) do you have to accomplish this milestone? Is this enough? If not, can you allocate additional resources?**

- **Plan: Create a detailed plan for accomplishing this milestone, being sure to include contingency plans in case the unexpected occurs.**

- **Calendar: Create a detailed calendar that tracks your progress and grades your results on a month-by-month basis.**

- **Partners: Who can you partner with to make accomplishing this milestone easier?**

- **Launch: When and how will you launch your efforts?**

- **Maintain: Are you prepared to maintain your efforts over the long haul to ensure successful completion of this milestone?**

- **Keep track: What specific measurements can you take on an ongoing basis to monitor your progress to determine if you are ahead, on track, or behind?**

- **Improve: How will you continually improve your efforts and results?**

Chapter 21

Secrets of Enlightened Savages

> *No pessimist ever discovered the secrets of the stars, or sailed to an uncharted land, or opened a new heaven to the human spirit.*
> Helen Keller

The 16 secrets you'll see in this chapter are what it ultimately means to be an enlightened savage. These secrets come with a performance guarantee: If you memorize all 16 secrets and run your life and/or business by the concepts they represent, then you will exceed your most optimistic projects, no matter how lofty. If you only memorize 15, you'll be in trouble, because you'll only have part of the puzzle. Remembering all 16 words is easy because they all end with the letters ENT. If you've read Jay Conrad Levinson's *Guerrilla Marketing* series, these words will sound very familiar, because all successful guerrilla marketers are enlightened savages at heart.

The following diagram shows all 16 of the secrets required to be an enlightened savage. The remainder of this chapter discusses each secret in more detail.

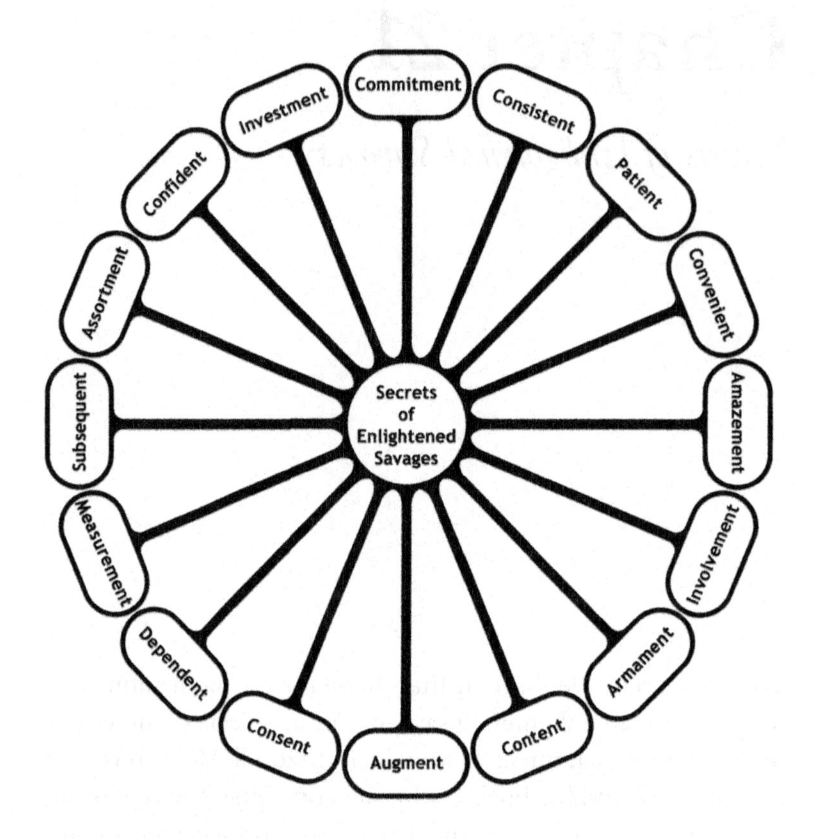

Commitment

All enlightened savages are 100% committed to themselves, their families, friends, associates, businesses, and the pursuit of living their dharma. This commitment motivates them to take action and to keep forging ahead, no matter what obstacles lie in their path. Ultimately, an enlightened savage's level of commitment is what will keep them moving through the uneasy period when they realize they've hit the walls of their comfort zone and are tempted to turn back.

> *There's no abiding success without commitment.*
> Anthony Robbins

If you are 100% committed to being an enlightened savage, then you will triumph in life and in business. Guaranteed.

> *If you want to be truly successful, invest in yourself to get the knowledge you need to find your unique factor. When you find it and focus on it and persevere, your success will blossom.*
> Sidney Madwed

Investment

Enlightened savages invest everything they have in everything they do, because they know that living one's dharma demands nothing less. Being anything less than totally invested in building the personal reality you want and need will either delay the success you want and deserve, or even eliminate it altogether. Your time on this planet is limited. How will you invest however much time you have left?

Consistent

You cannot perform in a manner inconsistent with the way you see yourself.
Zig Ziglar

Remember the fable of the hare and the tortoise? The rabbit ran ahead, then stopped and fell asleep under a tree. By the time he woke up, the slow, plodding tortoise had won the race. Enlightened savages know that steady, consistent progress is the best way to ensure the successful completion of everything they take on.

When planning efforts from major life's goals to simple daily tasks, enlightened savages think long term. They know that the best results take time, and allocate their resources accordingly.

Confident

If you are prepared, you will be confident, and will do the job.
Tom Landry

Remember that deciding to live your dharma means that you might need to sacrifice everything you have in order to get everything you need. An enlightened savage must therefore have a huge reservoir of confidence that will carry them through this transition and beyond. They know that they have the power to build any personal reality they want (within limits), and are confident in that power. They do this by letting go of attachments and making conscious choices about how to interpret everything that happens to them no matter what.

Patient

A patient mind is the best remedy for trouble.
Plaut

The best things in life take time to grow and mature, even without the many obstacles that might confront you. Enlightened savages know that they must never let life's many little fires rattle them. They also embrace the idea that everything happens when it should, and that forcing things never works. Further, enlightened savages know that not everyone shares their positive, life-affirming programming. They therefore exhibit infinite patience when dealing with the world and everyone in it.

Patience is not only a virtue; it is the secret to saving limited energy and efforts for the things that really matter when they matter.

Assortment

Variety is the condition of harmony.
Thomas Carlyle

Have you ever known anyone who always approaches any challenge and opportunity in the exact same way? It's like beating your head against the same old wall and expecting different results. Enlightened savages know that they need an assortment of tools and methods to accomplish their goals and tasks. They always tailor their approach to the situation at hand.

Convenient

If you want to make an easy job seem mighty hard, just keep putting off doing it.
Olin Miller

Never forget that work equals results, not effort. Enlightened savages always look for the easiest way to get results from everything they do. Always seeking ease and convenience removes the stress people feel when they make the error of defining work as effort. As an enlightened savage, you know that there is no such thing as hard work, just results that are either easy or hard to come by.

You might worry that always seeking convenience might lead to laziness and inaction. Worse, it could lead to exploiting people and resources. Fear not. If you are committed to being an enlightened savage, then you know that part of your success depends on the people around you, and will never feel tempted to take advantage of anyone.

Subsequent

The presence of a long-term, conscious goal has helped me maintain stability through the ubiquitous changes of over half a century.
Craig Mary

We live in an instant gratification society, where people expect their efforts to yield immediate results. They see life as a sort of giant vending machine, where results are dispensed as soon as the token is inserted. Enlightened savages know that life is a lot more like a fruit tree that requires ongoing care and nourishment to bear fruit. They understand that the moment they're ready to give up is the moment that their reward is almost within reach.

Plan for the long term; take your easy steps. You may not see your desired results immediately. Enjoy the journey, knowing that you will get where you're going at the end of (subsequent to) your travels.

Amazement

If you'll not settle for anything less than your best, you will be amazed at what you can accomplish in your lives.
Vince Lombardi

Enlightened savages are constantly amazed, both by the abundance surrounding them, and by their ability to build their desired personal reality. This state of childlike wonder spurs them to learn and experiment more, and in so doing find newer, easier ways to fulfill their life's mission. The generosity and support of others is another source of amazement to the enlightened savage, who never takes such kindness for granted. This state of amazement leaves enlightened savages open to experiencing joy and fulfillment, because they know and appreciate how precious life is and the importance of savoring every moment.

Measurement

One accurate measurement is worth a thousand expert opinions.
Admiral Grace Hopper

The best kind of progress is that which can be measured against the desired goal or outcome. Enlightened savages prepare and follow detailed plans that guide them along every step of every journey they take, making sure to reward themselves for each step. They see everything they do both for what it is and as part of the bigger picture of creating and living the life they are here to live. All enlightened savages measure success one easy step at a time.

In addition to measuring progress, enlightened savages measure their comfort. If they find themselves growing uncomfortable, they adjust the size of their steps accordingly. If their prey instincts are screaming at them to take cover, they slow their pace and maintain their rewards. If they are feeling bored by slow progress, they accelerate their plans. Doing this helps ensure that enlightened savages always feel comfortable and content.

Involvement

Growth means change and change involves risk, stepping from the known to the unknown.
George Shinn

There is no substitute for 100% involvement in one's own life and dharma. Enlightened savages are firmly in the driver's seat of their own lives and take full responsibility for realizing their lives. Detachment and withdrawal do not exist in the enlightened savage's personal reality.

Timing is key. The enlightened savage involves herself totally in what she is doing, then moves on when the step or task is complete without looking back. Likewise, she never involves herself too early, because she knows that the present demands all of her attention and focus.

Dependent

We are all dependent on one another, every soul of us on earth.
George Bernard Shaw

No person is an island, and enlightened savages are no exception. Everyone depends on many other people for the many necessities of daily life. Enlightened savages find people they can depend on, people who support their dharma and help in any way possible. This may mean giving up people who are unsupportive, something the enlightened savage never hesitates to do, because she knows that this lifetime is anything but permanent, and that this one chance might be all we get.

People also depend on enlightened savages. The enlightened savage knows this and does everything in her power to honor her commitments to those who depend on her. She never encourages or accepts dependency for its own sake; rather, she bases her decisions on finding and accomplishing her own dharma.

Armament

I do not believe you can do today's job with yesterday's methods and be in business tomorrow.
Nelson Jackson

Every enlightened savage arms her or himself with the tools and skills s/he needs to accomplish whatever task is at hand. She knows that having the proper tool for the proper job makes getting results far easier. Getting results with ease is the goal of every enlightened savage, meaning that finding and obtaining the correct tools always figures heavily in every plan they create.

Consent

Your life works to the degree you keep your agreements.
Werner Erhard

An enlightened savage never forces their will on anyone else. Everything they do that involves other people occurs with those people's full knowledge and consent. Where consent is not given, the enlightened savage finds another way of accomplishing whatever it is they are seeking to accomplish.

Obtaining consent builds consensus, which builds unity of purpose, which reduces conflict and sharpens focus. All of this combines to make obtaining the desired results faster and easier, the goal of every enlightened savage.

Content

Health is the greatest gift, contentment the greatest wealth, faithfulness the best relationship.
Buddha

All enlightened savages are supremely content with themselves and with their lives, because they know that they created everything they have and everything they are. If they are on the path to achieving their life's mission, they rejoice, for they know that their lives truly do have purpose. If they are seeking the right path, they rejoice, for they know that everything that isn't right serves to point them in the right direction. It's good to be the monarch; every enlightened savage is the undisputed monarch of her or his own life.

Content, in the form of actual meat or substance, is also extremely important to enlightened savages, because they know that smoke, mirrors, and obfuscation only confuse and hide the true path, and make achieving their goals that much more difficult.

Augment

We increase whatever we praise. The whole creation responds to praise, and is glad.
Charles Fillmore

Just because you are on the correct path and realizing your desired life and lifestyle does not mean you should rest on your laurels. Enlightened savages are constantly seeking new ways to augment and refine their efforts to make them easier and more effective. They look to employ everything they obtain and everything they learn for this one purpose.

Questions

The degree to which you master the 16 ENT words is the degree to which you will obtain the personal and business success you deserve. The good news is that you already know how to use every single one of these words. How? Think about the processes you've recently been through and about all the old destructive programming you've eliminated. You have already demonstrated and used each of these words, and are well on your way to becoming an enlightened savage.

Rate yourself from 1 (not at all) to 10 (perfectly) on how well you have mastered the secrets of enlightened savages:

- **Commitment** _____
- **Investment** _____
- **Consistent** _____
- **Confident** _____
- **Patient** _____
- **Assortment** _____
- **Convenient** _____
- **Subsequent** _____
- **Amazement** _____
- **Measurement** _____
- **Involvement** _____
- **Dependent** _____
- **Armament** _____
- **Consent** _____
- **Content** _____

- **Augment _____**

- **How can you improve your mastery of each of the secrets of enlightened savages where you scored less than a perfect 10?**

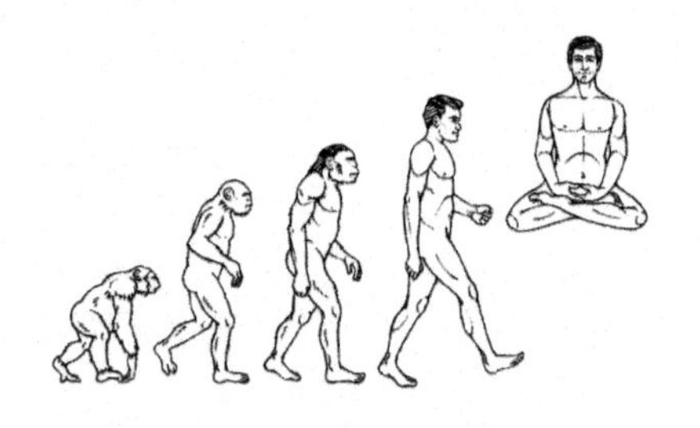

Chapter 22

Way of the Enlightened Savage

This chapter explains how enlightened savages view themselves, the world, and their actions. It's a vision of a life filled with purpose, accomplishment, and balance.

The Journey is the Goal

Some men give up their designs when they have almost reached the goal; While others, on the contrary, obtain a victory by exerting, at the last moment, more vigorous efforts than ever before.
Herodotus

The enlightened savage knows that the journey is the goal. She also realizes that she is in full control of how she sees her life and exercises that control to see the good in all things. If she is dissatisfied with her journey, it is because she is missing the point of the journey itself. Unlike the old-fashioned approach that often requires gigantic sacrifices for the sake of the goal, the enlightened savage places the goal of a pleasant journey ahead of the mere notion of sacrifice. In other words, she never engages in sacrifice for its own sake.

Obtain and Maintain Balance

In a balanced organization, working towards a common objective, there is success.
T.L. Scrutton

Far too many people place work ahead of leisure, showing no respect for their own personal freedom. An enlightened savage cherishes her freedom as much as her work, and achieves balance from the very start. She builds free time into her work schedule so that balance is part of her life. She respects her leisure time as much as her work time, never allowing too much of one to interfere with the other. Balance is easy for the enlightened savage to attain, because she knows that work is a measure of results, never of effort.

Pace Yourself

The enlightened savage is not in a hurry. A false need for speed frequently undermines even the best-conceived strategies. Haste makes waste and sacrifices quality. The enlightened savage is fully aware that patience is her ally. She uses intelligent planning to eliminate most emergencies that call for moving fast. Her pace is always steady, but never rushed.

To climb steep hills requires slow pace at first.
William Shakespeare

Eliminate Stress

An enlightened savage uses stress as a benchmark. If she feels any stress, she takes it as proof that she is moving in the wrong direction and/or doing things the wrong way. No enlightened savage ever accepts stress as part of doing business. On the contrary, she recognizes that any stress is a warning sign that something's wrong with her methods, plans, and/or life itself. She makes adjustments to eliminate the cause of the stress, rather than simply medicating the stress itself. You can do this by realigning core beliefs, finding and living your dharma, and staying the course.

Pressure and stress is the common cold of the psyche.
Andrew Denton

Enjoy Work

The enlightened savage looks forward to work. She has a love affair with her work and considers herself blessed to be paid for doing the work she does, no matter what that work is or how she get paid. She is good at her work, energizing her passion for it in a quest to learn more about it, and improve her understanding of it, thereby increasing her skills. The enlightened savage doesn't think about retirement, because she would never want to stop doing the work she loves and was born to do.

Losers live in the past. Winners learn from the past and enjoy working in the present toward the future.
Denis Waitley

Be Strong

Only strength can cooperate. Weakness can only beg.
Dwight D. Eisenhower

The enlightened savage has no weaknesses. She is effective in every aspect of her life, because she has filled in the gaps between her strengths and talents with people who are proficient at the skills she lacks. She is very much the team player, and teams up with enlightened savages who, like herself, share the team spirit and possess complementary skills. She values her teammates as much as old-fashioned entrepreneurs valued their independence.

Think Fusion

Joint undertakings stand a better chance when they benefit both sides.
Euripides

The enlightened savage is fusion-oriented in her life and business. She is always seeking opportunities to fuse her life and her business with other enlightened savages and enterprises in her town, nation, and the world.

She is willing to combine efforts, skills, information, leads, and anything else to increase her effectiveness and reach, while reducing the cost and struggle required to achieve her goals.

Don't Fool Yourself

The easiest person to deceive is one's own self.
Edward G. Bulwer-Lytton

The enlightened savage does not kid herself. She knows that overestimating her own abilities risks skimping on the quality she represents to her friends, family, coworkers, customers, employees, investors, partners, and, most of all, herself. She forces herself to face reality on a daily basis, and realizes that she must always evaluate everything she does in the glaring light of what is really happening, instead of what should be happening. When what is really happening differs from what should be happening, the enlightened savage takes active measures to create her desired personal reality and outcome, because she knows that she has the power to create any personal reality she wants (again, within limits).

Live for Today

The ability to be in the present moment is a major component of mental wellness.
Abraham H. Maslow

The enlightened savage lives in the present. She is well-aware of the past, and very enticed by the future, while residing firmly in the here and now. She embraces her knowledge and place in the present, leaving future ideas and goals on the horizon, right where they belong, until such time as they are ripe and ready. She is alert to the new, wary of the avant-garde, and only wooed from the old by improvement, not merely by change.

The enlightened savage understands the precious nature of time. She doesn't buy into the old lie that time is money, because she knows in her heart that time is far more important than money. Time is life. Everyone she meets feels the same way about time, so she never wastes anyone's time. She is the epitome of efficiency, but never lets it interfere with her effectiveness.

Follow a Plan

The enlightened savage always operates according to a plan. She knows where she is, where she is going, and how she will get there. She is prepared, knows that anything can and will happen, and can deal with the barriers to personal and business success, because her plan has foreseen the obstacles, and shows exactly how to surmount them. The enlightened savage reevaluates her plan regularly, and does not hesitate to make changes in it, though commitment to the plan is part of her very being. Knowing that progress is joyful, she rewards herself for every small step she takes in accomplishing any plan.

You return and again take the proper course, guided by what? By the picture in mind of the place you are headed for...
John McDonald

Be Flexible

The enlightened savage is flexible. She is guided by a strategy for success, and knows the difference between a guide and a master. When change becomes necessary, the enlightened savage makes that change because change is part of the status quo, and must never be ignored or fought. She can adapt to new situations, realizes that service is whatever her life's mission requires it to be, and knows that inflexible things become brittle and break.

Prepare yourself for the world, as the athletes used to do for their exercise; oil your mind and your manners, to give them the necessary suppleness and flexibility; strength alone will not do.
Lord Chesterfield

Seek Results

Work joyfully and peacefully, knowing that right thoughts and right efforts will inevitably bring about right results.
James Allen

The enlightened savage seeks results over growth. She focuses on profitability and balance, vitality and improvement, and value and quality over mere size and growth. Her plans call for steadily increasing profits without sacrificing personal time. Her actions are, therefore, oriented to hitting those targets instead of growing just for the sake of growth. She is wary of becoming large, and does not equate hugeness with excellence, because she knows that true profits come from finding and following her life's mission.

Be People-Oriented

The mechanics of industry is easy. The real engine is the people: Their motivation and direction.
Ken Gilbert

The enlightened savage depends on many people. The age of the lone wolf is over. The enlightened savage depends heavily on everyone around her and chooses these people to be part of her support structure. She never forgets that she got where she is with her own wings, her own determination, her own smarts, and, as an enlightened savage, with a little help from a lot of friends.

Never Stop Learning

Rich people without wisdom and learning are but sheep with golden fleeces.
Solon

The enlightened savage is constantly learning. A seagull flies in circles in the sky, looking for food in an endless quest. When it finally finds the food, the seagull lands, then eats its fill. When it has completed its meal, the seagull returns to the sky, only to fly in circles again, searching for food although it has already eaten. Humans have only one comparable instinct: the need for constant learning. Enlightened savages have this need in spades. The one thing enlightened savages know for certain is that they know nothing for certain.

Be Passionate

By believing passionately in something that still does not exist, we create it. The nonexistent is whatever we have not sufficiently desired.
Nikos Kazantzakis

The enlightened savage is passionate about life. Her enthusiasm for what she does is readily apparent to everyone who sees her life and life's work. This enthusiasm spreads to everyone who associates with her. In its purest form, this enthusiasm is best expressed as passion, an intense feeling that burns within her and reveals itself in the devotion she demonstrates towards her life and business.

Stay Focused

The enlightened savage remains focused on her goal. She knows that balance does not come easily, and that she must rid herself of her ancestors' values and expectations by upgrading her mental programming, removing destructive core beliefs, and remaining focused and steadfast on her journey. She sees the future clearly while concentrating upon the present, because she is aware that the minutiae of life and business can distract her. She therefore does whatever is necessary to make those distractions only momentary.

Focus 90% of your time on solutions and only 10% of your time on problems.
Anthony J. D'Angelo

Be Disciplined

The enlightened savage is disciplined about the tasks at hand. She is keenly aware that every task she places on her daily calendar is a promise she is making to herself. As an enlightened savage who does not kid herself, she keeps those promises, knowing that achieving each step of her goals is more than an adequate reward for her discipline. She finds it easy to be disciplined, because of the payback offered by the leisure that follows.

Discipline is the refining fire by which talent becomes ability.
Roy L. Smith

Be Organized

The enlightened savage is well-organized at home and at work. She knows that she must never waste valuable time looking for items that have been misplaced, and therefore organizes as she works, and as new work comes to her. Her sense of organization is fueled by the efficiency that results from it. Here again, balance is essential; while she is always organized, the enlightened savage never squanders precious time by over organizing.

An Englishman, even if he is alone, forms an orderly queue of one.
George Mikes

Be Upbeat

I became an optimist when I discovered that I wasn't going to win any more games by being anything else.
Earl Weaver

Finally, and most important, the enlightened savage has an upbeat attitude. She knows all too well that life is only as fair as she makes it, problems arise, to err is human, and the cool shall inherit the Earth. She takes obstacles in stride, keeping her perspective and sense of humor, and always being thankful for the step up. Her ever-present optimism is grounded in an ability to perceive the positive side of things, recognizing the negative, but never dwelling there. Her positivity is contagious.

Questions

Will you follow the way of the enlightened savage? Write down your answers to the following questions:

- **Do you agree that getting there really can be half the fun? Are you willing to have fun and associate joy with your life's journey?**
- **How will you seek and maintain balance in everything you do?**
- **What specific things can you do to pace yourself so as to maintain energy and resources for the long haul?**
- **How can you eliminate stress from your business and personal lives? To what extent has becoming an enlightened savage helped you with stress reduction?**
- **How can you enjoy your work more? Describe your ideas in as much detail as possible.**
- **How can you nurture your strengths while fortifying your weak points? Are there any methods you can employ to cope with your weak points that don't involve negative self-talk?**
- **How can you partner with other enlightened savages to create even more powerful results? Does anyone specific come to mind? If so, who and why?**
- **Are you willing to be totally honest with yourself so that you can always identify challenges and problems before they become overwhelming?**
- **Will you live for the moment knowing that the past is over and done with and that the future has not yet arrived?**
- **Do you have a plan for how to accomplish your goals? If not, why not? If so, are you 100% committed to executing that plan come what may? If not, why not?**

- Are you willing to be flexible so as to bend around challenges instead of letting them break you?

- Will you focus on the results in stead of on the effort? Do you own the fact that "work" equals output, not effort?

- Are you a people person? If not, how can you meet and get to know more people? Be as specific and detailed as possible.

- Are you committed to lifelong learning and growth?

- Will you follow your passions so as to live your dharma? If not, why not?

- How well focused are you? What can you do to improve your focus?

- How well disciplined are you? What can you do to improve your self-discipline?

- How well organized are you? What can you do to improve your organizational skills?

- Are you upbeat? How can you become even more of an optimist?

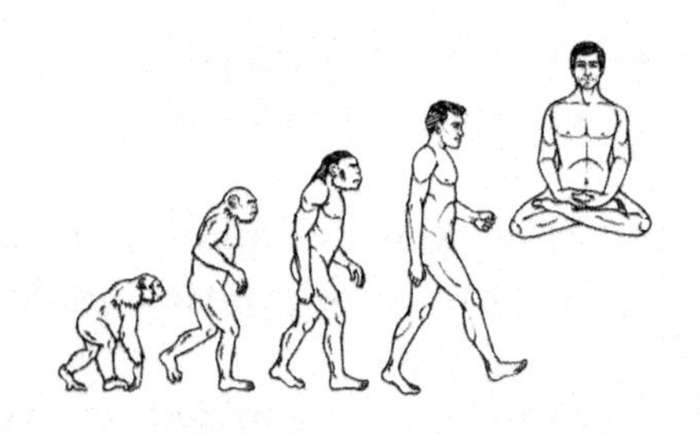

Chapter 23

The "Law of Attraction" is Wrong!

The Law of Attraction attracts to you everything you need, according to the nature of your thought life. Your environment and financial condition are the perfect reflection of your habitual thinking. Thought rules the world.
Joseph Edward Murphy

We all need to feel like we're in control of our lives and our environments. We like to believe that there is such a thing as security and that we can bend the forces of nature to our will. On the other hand. losing control is a very scary feeling. As I said in Chapter 15, I can't imagine a worse feeling for a prey animal than hopelessness. Enter the self-help industry where legions of experts peddle tens of thousands of products that promise to give you the ability to "attract" and/or manifest" anything you want, and I do mean anything—from money to relationships, health, and even freedom from aging. These claims are all based on the "Law of Attraction," which basically says that positive thinking, acting as if you already have what you seek, and exuding happiness and gratefulness is all you need to place the cosmic equivalent of a pizza delivery order that is absolutely guaranteed to arrive. Quantum physics and ancient wisdom are cited to "prove" that the "Law of Attraction" is real and reams of testimonials are trotted out to offer additional evidence.

There are some people who live in a dream world, and there are some who face reality; and then there are those who turn one into the other.
Douglas Everett

Does each of us "manifest" our own reality? Sort of. There is no such thing as objective reality because everything that we experience in our lives is filtered through our core beliefs long before we are aware of even having that experience. We tend to repeat the same old patterns because of our physical addictions to the cocktails of chemicals in our bloodstream that are released whenever we have an emotion. Can anybody build a different personal reality by following the steps in this book? Absolutely. That said, the entire process of creating that change involves small steps because the whole goal is to wean you off the emotional addictions that are preventing you from living the life you want and replacing them with new addictions that will open new doors for you. The human animal is literally a junkie because we depend on emotions driven by core beliefs to keep ourselves from being killed and eaten. The trick is to adjust that addiction by pumping our veins full of new emotional juice. The healthy way to do this is through the gradual processes described in this book. The unhealthy way is through abusing drugs.

It is not good enough for things to be planned. They still have to be done; for the intention to become a reality, energy has to be launched into operation.
Pir Vilayat Khan

Can you put out a wish/intention/affirmation and suddenly get what you want? Yes. The problem, though, is that the odds of actually doing this are extremely small. You hear stories of people experiencing dramatic changes all the time, which seems to lend credence to the "Law of Attraction." What you don't hear is that with billions of people on the planet, some are bound to hit the cosmic jackpot just like some are bound to win the lottery. That does not mean that buying a lottery ticket is ever a smart investment.

A Word to the Faithful

This chapter will not be easy reading if you believe in the "Law of Attraction," especially if you are part of the self-help industry with products based on the "Law of Attraction." You may think I am wrong, and with good reason, because so many of the people you know and trust believe in this "law." You may also think that I am attacking you and/or your beliefs. Nothing could be further from the truth. On the contrary, it is my sincere hope that reading and understanding this chapter will replace your belief in the "Law of Attraction" with the truth, which will make everything you know and assume based on this "law" both more subtle and exponentially more powerful. Thus, the more you think you should stop reading this chapter, the more you should read it with an open mind. I guarantee that you will be very glad you did.

Confusing the words 'wish,' 'faith' and 'pray' with each other usually just results in a minor grammatical faux pas, but when any of these words, especially 'hope,' is confused with 'action,' the results are much more devastating.
Bo Bennett

As I explain in *The Divine Savage*, "I believe that most self-help experts who subscribe to the 'Law of Attraction' do so in good faith precisely because of the *truthiness* (appearance of being true) and the overwhelmingly positive message contained therein that essentially says, 'If you can dream it, you can do it.' A strict reading of quantum mechanics where a waveform of possibilities collapses into actuality does mean that literally anything is possible and that the 'Law of Attraction' is technically correct. The problem is that this correctness only applies under very esoteric conditions, which effectively renders the 'Law of Attraction' null and void."

Great feelings will often take the aspect of error, and great faith the aspect of illusion.
George Eliot

Let's find out why.

The Quantum Waveform

Do not expect to arrive at certainty in every subject which you pursue. There are a hundred things wherein we mortals... must be content with probability, where our best light and reasoning will reach no farther
Isaac Watts

Set up as many identical quantum experiments and starting conditions as you like, and you would be forgiven for expecting the same results every time. Without this consistency, science itself would grind to a halt. Students could not be taught, and discoveries could not be validated through replication. This works well in the *macro* world, the world we are accustomed to seeing, hearing, feeling, smelling, and tasting; however, the quantum world is a very strange place, because repeating the same experiment with the same starting conditions can indeed yield different results.

This amazing truth exists because all objects have a *probability wave* (also called a *waveform*) that measures the probability of finding that object at any given location in spacetime. We cannot know this location until we observe it. The particle exists in a *superposition* of all possible states and locations until the observation/experiment occurs. What we think of as solid bits of matter are actually waves that extend throughout the universe, unless and until we measure it. Before a measurement, all we can do is predict possibilities. If we had all of the information about a particle, we could predict the exact results, but *uncertainty* (a natural law discovered by Werner von Heisenberg and called the *Heisenberg Uncertainty Principle*) prevents us from ever having all of the information or the resources to crunch all of the numbers. We are forever limited to educated guesses.

The waveform itself represents the probability of finding an object in any location within the wave. The height of the wave at any point represents the probability of finding the object at that exact location. This is not the same as saying that, for example, my son Logan has a 99% or higher chance of being at school during school hours on any given day. My son is either at school during that time or he isn't; however, for a quantum object, the waveform is the same as the object itself. The waveform of an atom is the atom itself. Looking for my son at school does not cause him to be at school; he is either there or not independently of my observation. Or is he? In the quantum world, observing a particle at a given location causes it to be there, and observing a particle not in a given location causes it to not be there. Finding the object instantly collapses the waveform to 1 (100%) in that location and 0 (0%) in all other locations. The pre-observation waveform only tells us the probability of what we will observe. We can only know where something is when we are staring right at it.

The laws of probability, so true in general, so fallacious in particular.
Edward Gibbon

Logan should be at school right now. The GPS tracking feature on his cell phone indicates that he is within 40 yards of his classroom. He therefore has a waveform that is 80 yards in diameter whose highest peak is centered on his desk in his classroom. Going to his school and looking in the classroom is the only way for me to collapse his waveform, which will then spread right back out as soon as my back is turned. His waveform could also be collapsed by someone else observing him in his classroom, which collapses the waveform for everyone else who looks through the window immediately afterward. Reporting that information to me (such as by calling me) would thus collapse my view of his waveform.

Logan's teacher is normally the first person to see him in his classroom. Believe it or not, we can therefore conclude that Logan is not in his classroom before his teacher observes him. This example applies equally to the quantum world: There are no actual atoms in any actual place unless and until we look for them, a fact that seems hard to swallow since each of us is looking at collections of countless atoms that form the objects we are used to seeing and that don't seem to be going anywhere.

The odds of hitting your target go up dramatically when you aim at it.
Mal Pancoast

Physicist Max Born saw the wave, not as a particle smeared across the universe, but simply as a probability function that tells us the likelihood of finding a particle in one location or another. This seems simple enough, except that it becomes meaningless to ask where the object is before an observation is made. Can we then distinguish between wave and object? Does the object have a position when we are not looking at it? Does the moon exist when nobody is looking at it? The short answer is, "Quite probably, but not definitely." The more precise answer is that the question itself is meaningless.

We all knew there was just one way to improve our odds for survival: train, train, train. Sometimes, if your training is properly intense it will kill you. More often—much, much more often—it will save your life.
Richard Marcinko

A probability wave for a macro object is large in a given area but quickly drops to near zero with distance. In this example, Logan's 80-yard waveform peaks above his desk in his classroom, shrinks rapidly to near zero by the time it reaches the edges of his classroom, and finally drops to zero 40 yards out. I know that it drops to zero because I made an indirect observation using the GPS locator. Absent this observation, my son's waveform would extend across the entire universe. The same thing happens with quantum objects, meaning that the odds of one particle tunneling any noticeable distance is extremely small. Multiply this by the number of particles involved (see Chapter 15 for an example), and you will soon discover why you need not worry about suddenly finding yourself somewhere else. It also explains why the "Law of Attraction" won't make you rich, in love, or healthy any time in the foreseeable future.

A Brief History

If past history was all there was to the game, the richest people would be librarians.
Warren Buffett

The term "Law of Attraction" appeared in the 1906 book *Thought Vibration* by William Walker Atkinson. It appeared again in the 1907 title *Prosperity Through Thought Force* by Bruce MacLelland claimed that, "You are what you think, not what you think you are." In 1910, *The Science of Getting Rich* by Wallace D. Wattles uses his interpretation of Hindu idealism to claim that believing in and focusing on what you want will realize—or manifest—that goal, while negative thoughts will of course manifest negative results.

Napoleon Hill (1883-1970) was one of the pioneers of the modern self-help industry. His 1928 book, *The Law of Success in 16 Lessons* contained many direct references to the "Law of Attraction" as did *Think and Grow Rich*, which was first published in 1937 and has sold more than 60 million copies to date. Hill explains that one must control both one's thoughts and the energy behind those thoughts to attract other thoughts and ultimate success. This beginning of this book features a teaser that talks about a "success secret" along with the promise to reveal portions of this secret at least once per chapter thereafter. Hill never spills all the beans, but does explain that "attraction" has something to do with it.

Other authors took this concept and ran with it but the "Law of Attraction" had to wait until the bestselling 2006 book and movie *The Secret* by Rhonda Byrne to reach the mainstream zeitgeist. Hundreds of thousands of copies sold within hours, with millions more selling over the succeeding years. Dozens—if not hundreds—of self-help experts have also jumped on the bandwagon. Countless seminars, books, movies, etc. have been sprouting up ever since. These days, one is hard pressed to find a system that is not based on "The Law of Attraction" as popularized in *The Secret.* If nothing else, *The Secret* has been a marketing coup; however, just because millions of people believe something does not make it true. For example, millions of people used to believe that the world is flat. The few who cling to that belief today are widely—and rightly—scorned.

Let us resolve to be masters, not the victims, of our history, controlling our own destiny without giving way to blind suspicions and emotions.
John Fitzgerald Kennedy

Let's take a look under the hood of *The Secret* to see what it actually says, after which I will mention some of the criticism raised about some of its claims as I reveal why the "Law of Attraction" is functionally useless. From there, I will conclude this Appendix by explaining why the Law of Collapse (or Law of Realization, if you prefer) is the *real* secret.

Inside The Secret

The Secret claims that historical luminaries such as, "Plato, Shakespeare, Newton, Hugo, Beethoven, Lincoln, Emerson, Edison, [and] Einstein" guard the ultimate secret to life, love, wealth, anything one could ever want—a statement that has eye-rolling parallels with the Holy Grail conspiracy that powers Dan Brown's very enjoyable *The Da Vinci Code.* The key difference is that *The Secret* says that this secret—the "Law of Attraction"—is quite real. Ms. Byrne describes an astoundingly fortuitous search for the "Law of Attraction" through the pages of history and her success in discovering the "modern day practitioners" of this ancient art. The rest of the book consists of quotes by these "practitioners" with snippets from

The charm of history and its enigmatic lesson consist in the fact that, from age to age, nothing changes and yet everything is completely different.
Aldous Huxley

Ms. Byrne thrown in to explain/add to some of these quotes, which include:

- **Bob Proctor:** The Secret gives you anything you want: happiness, health, and wealth.
- **Joe Vitale:** You can have, do, or be anything you want.
- **John Assaraf:** We can have whatever it is that we choose. I don't care how big it is.

I can almost see Mike Myers as Austin Powers in *The Spy Who Shagged Me* retorting to Mr. Assaraf with, "Well, I want a toilet made of solid gold, but it's just not in the cards now, is it?" Believe it or not, that response goes to the heart of why the "Law of Attraction" is useless—but I'm getting ahead of myself.

The "Law of Attraction" Explained

The secret of success is constancy of purpose.
Benjamin Disraeli

We all know that the universe operates by a set of natural laws and forces with extremely precise values. We also know that life as we understand it would be impossible if some of these laws or values were even slightly different. (This is called the *anthropic principle.*) Of these, the "Law of Attraction" is by far the most powerful. It takes center stage because we are attracting absolutely everything we have in our lives for good or for ill. As Bob Proctor explains it, "Whatever is going on in your mind you are attracting to you. Every thought of yours is a real thing—a force." All successful people rely on The Secret either consciously or unconsciously by thinking about abundance and wealth, while deftly eschewing and avoiding any negative thoughts that could bring the entire house of cards down around their ears. People who go through cycles of rags and riches use The Secret sporadically, which explains their ups and downs.

The secret of getting ahead is getting started.
Sally Berger

At its root, the "Law of Attraction" is a magnet, and everything in our lives is metal. Thinking about something we really want brings that thing into our lives. Thoughts have frequencies like individual TV stations that radiate out into the universe like messengers carrying a cosmic dinner delivery order. This order will always be delivered piping hot without any possibility of exception. Change the frequency, and you could end up ordering Chinese, which could really ruin your day if you were hoping for pizza. Thinking, "I can taste the cheesy pepperoni goodness and feel my happy tummy!" is a surefire way to get pizza while thinking, "I really don't want Chinese food!" is the best possible way to guarantee a piping hot plate of chow mein in your imminent future. In other words, focusing on what you want as if you already had it is The Secret; thinking about things you don't want brings those undesirables into your life because the universe only hears "Chinese" or "debt" while skipping over the "I really don't want..." or "I need to get out of..." parts, respectively. It's the subject of the thought that counts, not the context. It's that simple.

Idealist Roots

The Secret says that, "Physicists tell us that the entire Universe emerged from thought!" Well, sort of. Some physicists, philosophers, and spiritual traditions do indeed tell us this that matter is a myth—that consciousness is the ultimate foundation of the universe, which creates the illusion of matter and separations such as the illusory distinction between "you" and "me." This model of reality is called *idealism.* Think of it as mind creates brain. Most mainstream science believes in *materialism*, which basically says that brain creates mind. The Judeo-Christian-Islamic model of reality is based on *dualism*, which says that both brain and mind are separate but equal. (Dualism is false because it violates the First Law of Thermodynamics dealing with conservation of mass and energy, leaving us with idealism or materialism.)

The secret of getting things done is to act!
Benjamin O. David

For reasons that require an entire book (*The Divine Savage*) to explore, I am a firm believer in idealism. There is ample evidence to safely conclude that consciousness and the thoughts that consciousness generates are indeed the stuff of which the universe is made. Does this mean that, "Nothing can come into your experience unless you summon it through persistent thoughts." as *The Secret* claims? Sort of. More on this later.

The Power of Positivism

If *The Secret* is to be believed, it is impossible to have negative thoughts while simultaneously feeling good because feelings are the universe's way of telling you what you are thinking. Bad

The secret to success is to start from scratch and keep on scratching.
Dennis Green

feelings are warnings to get happy right quick lest bad things be manifested.

- **Bernard Beckwith:** You can begin right now to feel healthy. You can begin to feel prosperous. You can begin to feel the love that's surrounding you, even if it's not there. And what will happen is the universe will correspond to the nature of your song. The universe will correspond to the nature of that inner feeling and manifest, because that's the way you feel.
- **Lisa Nichols:** Your thoughts and your feelings create your life. It will always be that way. Guaranteed!

The Secret asserts that the most powerful positive emotion is love—an assertion I absolutely agree with (see Chapter 15).

Placing Your Cosmic Pizza Order

The secret of happiness is to admire without desiring.
Francis H. Bradley

According to *The Secret*, the universe is the ultimate waiter on wheels who shares some key traits with Aladdin's genie. All Aladdin has to do is rub his magic lamp to have the genie therein at his beck and call. To continue the pizza example, if Aladdin—or anyone for that matter—wants pizza, all s/he need do is ask, which alerts the universe to your wishes and sets the ethereal cosmic chefs into motion. Joe Vitale says this process is, "really fun. It's like having the Universe as your catalogue. You flip through it and say, 'I'd like to have this experience and I'd like to have that product and I'd like to have a person like that.' It is you placing your order with the Universe. It's really that easy."

The secret of staying young is to live honestly, eat slowly, and lie about your age.
Lucille Ball

Placing the order—once will do, no need to keep asking—is step one. Step two is to believe that whatever you want is already in hand with every fiber of your being whether you can see it or not. After all, one need not perseverate after ordering that pizza; one can simply relax, safe in the knowledge that the delivery process is proceeding towards its inevitable successful conclusion. How these practitioners explain mistaken or misrouted pizzas or other "orders" is not discussed. I, for one, assume that there is a very high chance that my pizzas will arrive as expected, but this does not always happen; am I "attracting" the wrong pizzas from time to time because I busy myself with other things and don't give the pizza much thought until the doorbell rings?

Step three is to actually receive what you ordered, which you have dreamed/visualized/seen yourself receiving, while letting yourself feel gratitude and happiness. This, of course, is a key ingredient in manifesting another pizza, beach house, yacht, sultry blonde, or whatever you fancy. Did I mention the happiness and gratitude?

In all chaos there is a cosmos, in all disorder a secret order.
Carl Jung

Above all, get yourself into the good "frequency" through affirmations such as, "I have the most gorgeous blonde girlfriend who loves me," "I have that pizza in my tummy right now," "I have a million dollars," etc. Facilitate this effort by test driving the car you want, viewing the house you want, anything it takes to make you feel like it's already yours. The more you do this, the more likely you are to receive it (if the car dealership and realtor don't get wise to you first and thwart your efforts by using their own affirmations to ensure that you will actually buy the goods.) Keep it up, and one day there it will be, just as easy as that!

When the doorbell rings, you must get up to go answer it to get the pizza or welcome the blonde into your love nest. In other words, mere thinking and affirmations are not enough; you will actually need to do something at some point, but don't worry: It won't feel like work.

The local pizza parlor may promise delivery within 30 minutes or a free pie, but the universe has no such guarantees because time itself is an illusion. *The Secret* says that, "Quantum physicists and Einstein tell us is that everything is happening simultaneously." and indeed they do—if and only if you are a photon traveling at light speed! Anything moving slower than light speed (which is everything that is not a photon) experiences a flow of time. Time itself may be an illusion—there is good reason to think so—but saying that, "Any time delay you experience is due to your delay in getting to the place of believing, knowing, and feeling that you already have it." flies in the face of common sense. Your pizza, girlfriend, Swiss bank account, yacht, etc. are out there... but you need to get there first before you can have them. In other words, Veruca Salt from *Willy Wonka & the Chocolate Factory* is absolutely correct to want everything NOW, but that does not mean she will get her way every time or even some of the time.

To produce things and to rear them, to produce, but not to take possession of them, to act, but not to rely on one's own ability, to lead them, but not to master them—This is called profound and secret virtue.
Lao Tzu

Miraculous Possibilities

Miracles happen to those who believe in them. Otherwise why does not the Virgin Mary appear to Lamaists, Mohammedans, or Hindus who have never heard of her?
Bernard Berenson

What can the "Law of Attraction" do for you? The possibilities are endless, so long as you focus on what you do want and not on what you don't want, because the universe does not take no for an answer. Want to lose weight? Visualize yourself at your ideal weight. Buy smaller clothes. Don't think about the weight you need to lose lest it stick with you. After all, "Food cannot cause you to put on weight, unless you think it can." Pizza and ice cream for everyone! Why not, especially when, "A person cannot think 'thin thoughts' and be fat. It completely defies the "Law of Attraction." More on this later.

Feeling a bit under the weather? Got cancer or any other disease/condition? Blame the bacteria or virus all you want, but the real culprit is stress. Disease is a feedback loop for stress, which can be countered by—wait for it—love and gratitude and visualizing yourself perfectly healthy. *The Secret* even includes a testimonial from a woman who used the "Law of Attraction" to cure breast cancer. Bob Proctor says that, "Disease cannot live in a body that's in a healthy emotional state." And to think my partner Jennifer actually wasted ten years of her life on premed, medical school, internship, and residency!

The Secret does correctly point out that our entire bodies are renewed every few years. So how can we possibly degenerate or have illness remain in our bodies? Apparently, the *Hayflick limit* (the number of times a cell reproduces before the regeneration stops) is just as emotionally dependent as one's bank account or love life. Who knew? I mean, besides Plato, Shakespeare, Newton, etc.

If *The Secret* is to be believed, the "Law of Attraction" can even be employed to, as *Harry Potter and the Sorcerer's Stone* put it, "put a stopper in death." John Assaraf says that, "Beliefs about aging are all in our minds. Release those thoughts and focus on eternal youth." On his Web site, Mr. Assaraf mentions breaking down in tears when he realized that his son was using The Secret and mastering his own life. I can only wonder whether that boy has shown any signs of growing/aging since then—but I think I know the answer.

Ingenuity, plus courage, plus work, equals miracles.
Bob Richards

I have seen the placebo effect in action on my own son. I am aware of studies proving that patient attitude plays a huge role in medical outcomes. In Chapters 2 through 4, I explained the prey instinct in humans, how it shapes our species's entire mentality and worldview, and how our fear of getting literally killed and eaten drives most of what we do. I know for a fact that our beliefs, emotions, and thoughts greatly affect our health. I know that negative emotions release toxins into the bloodstream. That said, one could argue that Buddhist monks are the most highly trained "Law of Attraction" practitioners on the planet because of their lifelong focus. There are many reports of dead monks taking a very long time to begin decomposing, which raises many interesting questions. Still, I have yet to learn of an immortal monk—or anyone, for that matter.

Expect checks in your mailbox instead of piles of bills. When a bill does arrive, see it as a check for ten times the amount due. Whatever you do, don't think about the debt you are in. Expect a check, and a check will arrive. Want that Maserati but the bank account won't quite cover it? Forget it, because you can afford it! The happier you are, the richer you will be. Oh, and don't worry: There is plenty for everybody.

Want a new relationship or to fix up an existing one? Here again, be happy. Focus on having that relationship with that lusty busty blonde and concentrate on her good attributes. The more you love yourself, the more you will find loving people.

The Power of Gratitude

The happier and more grateful you are, the happier people you will attract. You get the idea. (I very much agree with this last point from personal experience, albeit not for the reasons offered by *The Secret.)* Make a list of everything you are grateful for. It is impossible to have more than you have now if you are not grateful for what you do have. All ungrateful thoughts carry that negativity to the universe, which responds with bad feelings to alert you to the imminent danger of additional loss. Don't stop there: Be grateful for the things you are "attracting" as well. Make sure to think happy, grateful thoughts as you drift off to sleep each night and continually find ways to work that into your life and habits.

The Law of Attraction attracts to you everything you need, according to the nature of your thought life. Your environment and financial condition are the perfect reflection of your habitual thinking. Thought rules the world.
Joseph Edward Murphy

Always Look on the Bright Side

Self-sacrifice is the real miracle out of which all the reported miracles grow.
Ralph Waldo Emerson

Never fight against what you don't want. Fight for what you do want. The example in *The Secret* involves anti-war sentiment and the admonition to advocate for peace instead of against war. I'll buy this because working toward a goal tends to be much easier and more rewarding than struggling against something.

The Universal Mind

Develop an attitude of gratitude, and give thanks for everything that happens to you, knowing that every step forward is a step toward achieving something bigger and better than your current situation.
Brian Tracy

According to *The Secret*, the universe consists of the Universal Mind, which "is the attractive force which brings electrons together by the 'Law of Attraction' so they form atoms; the atoms bought together to create molecules, [etc.]" John Hagelin says that, "Quantum mechanics and quantum cosmology confirm that the universe essentially emerges from thought and that all of the matter around us is just precipitated thought." This is idealism at its most fundamental, which has been all but conclusively proven, as far as I am concerned (see *The Divine Savage*). Bernard Beckwith agrees, saying that, "Are there any limits here? No, we are unlimited beings. We have no ceiling, The capabilities and he gifts and the power that is within ever single individual on the planet is unlimited."

At face value, the "Law of Attraction" seems compatible with everything we have discussed in this book, my tongue-in-cheek descriptions notwithstanding. We may know we live in a universe built on consciousness, that matter is an illusion, etc. We certainly know from firsthand experience that attitude is everything. This book stresses getting rid of negative beliefs and emotions to open you up to achieving your goals. Given all of this, how the "Law of Attraction" be as useless as I say it is? You are about to find out.

The Flaws in the "Law"

Every time you go out on the ice, there are slight flaws. You can always think of something you should have done better. These are the things you must work on.
Dorothy Hamill

If you have come this far and still believe in the "Law of Attraction," fine. It is my hope that the next few pages will replace that belief with a much more powerful truth. If your belief in the "law" has been shattered, don't worry—all will be as well as it can be. If you never believed in the "law," that's also OK; just keep your mind open for what is to come.

The ABC News article *Science Behind 'The Secret'?* says that, "Critics say *The Secret* is not only wrong, it is dangerous, leading people to believe you can get what you want, whether it's getting rich, curing disease, losing weight... simply by thinking positively about it." About the claimed medical benefits of the "Law of Attraction," this article quotes Dr. Richard Wender, president of the American Cancer Society as saying, "I want to be very clear that there is no evidence that people attract cancer by their thoughts." and, "If some person chose to strictly follow the steps in this book, there is a risk that they could die needlessly." Physicist Brian Greene is quoted as saying, "If by 'Law of Attraction,' they have this notion of having a thought and it attracts like thoughts, I can assure you that quantum mechanics has nothing to say about that." As for historical luminaries guarding secret knowledge, Greene says, "Look, I've never met any of those guys, but I have zero evidence that any of them would've held on to any fundamental secret about the world and not shared it."

I have full cause of weeping, but this heart shall break into a hundred thousand flaws or ere I'll weep.
William Shakespeare

These criticisms are damning in their correctness for reasons I am about to share. As you read, please keep in mind that any one criticism is enough to sink the "Law of Attraction." Taken all together, the only conclusion we can reasonably make is that this "law" is technically real, but practically little more than empty wishful thinking.

Creating our Personal Reality

We know from Chapter 4 that humans build our personal realities using the very predictable ETEAR cycle: Our earliest emotions (E, that probably begin before birth) create thoughts that form beliefs (T). These thoughts/beliefs drive our emotions (E), the addiction to which spurs us to take action (A) to manufacture whatever result we are familiar with (R). Humans are fundamentally emotional creatures; our logic exists to rationalize and validate our emotions and keep us pumping the same addictive chemical cocktails (what neuroscientist Candace Pert calls the "molecules of emotion") through our veins. This process fuels more emotions and around and around we go.

The mysteries of faith are degraded if they are made into an object of affirmation and negation, when in reality they should be an object of contemplation.
Simone Weil

The assertion that emotions warn us about bad thoughts and act as "manifestation warnings" is wrong, because it assumes that humans are logic-driven, when we already know that we are emotionally driven. Conscious thought can bypass our emotions and induce us to take different actions, which triggers a withdrawal process that is every bit as real as withdrawing from heroin or other such drug. The real secret is therefore to take action and to persevere despite all obstacles, being sure to reward yourself copiously for taking these steps—the end goal being to replace your current emotional addictions with new ones. As emotional prey animals, we can never escape the fear of being killed and eaten and will always follow our survival instincts to the letter. We can, however, reprogram those instincts.

Inherent Contradictions

Contradiction is not a sign of falsity, nor the lack of contradiction a sign of truth.
Blaise Pascal

The quotes I presented above—all of them lifted verbatim from the very pages of *The Secret* itself—say that we can do, have, and be absolutely anything we want, if we follow the three steps of placing the order, visualizing having that order fulfilled, and being happy and grateful no matter what. The "Universal Mind" is *omnipresent* (everywhere at once), *omniscient* (all knowing), and *omnipotent* (all powerful), and cannot refuse our every wish. The concept of universal mind/Godhead is absolutely correct for reasons I explain in *The Divine Savage*, as is the assertion that the universe follows extremely precise laws to the last iota. *The Secret* quotes Bernard Beckwith saying that, "We live in a universe in which there are laws..." So far, so good, except that Beckwith goes on to say, "... just as there is a law of gravity. If you fall off a building it doesn't matter if you're a good person or a bad person, you're going to hit the ground."

I believe that truth has only one face: that of a violent contradiction.
Georges Bataille

If we can have anything we want to the extent of defying aging and death, then it seems only reasonable to assume that we should be able to wish ourselves a feather-soft landing after stepping off a high balcony. Why can the universe bend its rules to give you anything you want while at the same time being incapable of bending those rules to give you anything you want. This one sentence alone dooms *The Secret* and the "Law of Attraction." I am flabbergasted that the editors let this one in, because most philosophies don't include the seeds of their own destruction quite so blatantly.

The universe must obey its own laws, because making one exception risks making another and another, until the very fabric of spacetime that makes life possible quite literally disintegrates. A universe without the laws ours has is incompatible with life. A universe that is self-consistent cannot possibly break its own laws. The extent to which the *anthropic principle* (the idea that the universe is finely tuned to allow life to evolve) applies, and to which sciences like quantum mechanics are as precise as they are, makes this self-evident. The universe is rational to its core. You cannot have your cake and eat it too, except under very precise circumstances governed by the quantum waveform, in which all things are indeed possible, but some things are far more likely than others.

There is no such uncertainty as a sure thing.
Robert Burns

Don't jump off any tall buildings. If you get sick, seek medical help. Accept the certainty that your life is going to end, and (most probably) continue to another form of existence in which you will be at once not yourself, and yet far more "you" than you are now. If you are overweight, eat less and exercise more. If you want more money, take steps to make more money. If you want a new relationship, go out and look for one. If you want to fix an existing relationship, take steps to fix it. In other words, use the laws of nature to your advantage. If any expert suggests that you can violate the laws of human nature or physics, the only questions you need to ask pertain to that expert's true knowledge, motives, and/or willingness to demonstrate their claims, period.

Heisenberg Would Not Approve

You already know that the Heisenberg Uncertainty Principle limits the extent of knowledge we can ever have, because there is an element of randomness built right into the foundations of the universe. We cannot avoid dealing with Heisenberg randomness. The assertion that we can have anything we want simply by "manifesting" is therefore false on its face. The "Law of Attraction" is again utterly falsified for all practical purposes.

Truth is confirmed by inspection and delay; falsehood by haste and uncertainty.
Tacitus

Shared Reality

Physicist Eugene Wigner threw a massive wrench into the idea that we can each manifest everything we want by pointing out that we are not alone. (This is called the *Wigner's friend* paradox). If two or more people observe the same waveform and

The Law of Attraction attracts to you everything you need, according to the nature of your thought life. Your environment and financial condition are the perfect reflection of your habitual thinking. Thought rules the world.
Joseph Edward Murphy

wish for different results, what happens? Physicist Amit Goswami explains the problem in his article *Consciousness and Quantum Physics* by saying, "Imagine that Wigner is approaching a quantum traffic light with two possibilities, red and green. At the same time, his friend is approaching the same light from the perpendicular road. Being busy Americans, they both choose green. Unfortunately, their choices are contradictory; if both choices materialize at the same time, there would be pandemonium. Obviously, only one of their choices counts, but whose?"

The answer that both Wigner and his friend get green lights half the time. This is accurate, in conformance with the quantum waveform, and a direct falsification of the "Law of Attraction." It is simply impossible for everyone to get everything they want all the time, period. We could not have an orderly universe that is fit for life any other way. Even if you believe in idealism where a single consciousness forms the universe itself, that consciousness cannot and will not contradict itself, no matter how hard its illusory subdivisions that we call "you" and "me" try. The only way anyone could get absolutely everything s/he wants would be if s/he were utterly alone in the universe—which kind of puts a crimp in any wish for a relationship!

One can make the argument that different people have different levels of willpower, and thus different levels of ability to make things go their way. One can then say that someone with perfect will can indeed have anything s/he wants. This makes perfect sense. Now find me just one person with perfect will. Each of us can indeed change what is possible in our lives and can make dramatic changes in a very short time; however, that is—again—a very far cry from "manifesting" your desires. Yes, you can and do build your own personal reality, and yes, you can change that reality. You just have to do so in a way that makes the best use of the laws of physics. Choosing to see an event as positive instead of negative using the conscious observer technique we discussed in Chapter 15 is absolutely not the same as manifesting a yacht just by wishing for it.

People may doubt what you say, but they will believe what you do.
Lewis Cass

The Quantum Tsunami

Let's recap: Quantum mechanics demonstrates the equivalence of energy and matter, and that all things exist in a state of superposition that encompasses all possible outcomes at once. This assortment of possibilities is contained within a quantum waveform that literally extends across the entire universe, and that literally never shrinks to zero. This explains both how you may suddenly find yourself orbiting Pluto or how your flat tire may spontaneously reinflate, and why you should absolutely not hold your breath waiting for either event—or anything similar—to happen.

The winds and waves are always on the side of the ablest navigators.
Edward Gibbon

As I explained above, my son's waveform spreads across the entire universe the moment he leaves my sight every weekday morning to go to school. The moment the door closes behind him, he could quite literally be anywhere in the universe. That said, his waveform—the probability wave that predicts the odds of him being in a given location at any given time—is extremely high along the route to his school before settling over his classroom when he arrives. There is an extremely high probability that chasing after him or visiting his school will find him exactly where he is supposed to be. The GPS tracker in his cell phone allows me to collapse his waveform any time I want. If the tracking feature says he is within a 40-yard radius of his classroom, then I know that his waveform is extremely high above his seat, much lower around the edges of his classroom, lower still in adjoining areas, and 0 at any distance more than 40 yards from his indicated location. My measurement of his location has collapsed his waveform to that 40-yard radius and my walking into his classroom would collapse his waveform to distance much less than the size of an atom. His waveform instantly spreads back out across the universe the moment the measurement ends, but always remains extremely high over his school while the day is in session.

Now is the season for sailing; for already the chattering swallow is come and the pleasant west wind; the meadows bloom and the sea, tossed up with waves and rough blasts, has sunk to silence. Weigh thine anchors and unloose thy hawsers, O Mariner, and sail with all thy canvas set.
Leonidas of Tarentum

The fact that the waveform is never zero between measurements means that, yes, you can manifest anything you want. The "Law of Attraction" is fundamentally correct in this assertion. You can indeed manifest anything you want from money to health, love, immortality, the ability to leap tall buildings in a single bound, anything at all. I could will my son to Pluto using the methods contained in *The Secret*, and it could happen. The only catch is that the odds are extremely low—so low that I could wait countless trillions of years before Logan's waveform randomly fluctuates to make such a thing statistically possible. Even the seeming exceptions to this rule such as the woman who cured herself of cancer by "attracting" health only prove the rule.

Here is a real-world example: Millions of people have visited Lourdes, France in hopes of receiving a miraculous cure. The total number of people who have actually been cured after visiting Lourdes is about 66. Dividing these 66 people by the millions of unfulfilled hopefuls reveals that the rate of cures attributable to that pilgrimage is many times lower than the statistical odds of spontaneous remission. Miracle? No. Attraction? Hardly. Quantum waveform and uncertainty (read: raw statics)? Absolutely. The quantum waveform is therefore a tsunami that utterly wipes out the "Law of Attraction."

It's just a job. Grass grows, birds fly, waves pound the sand. I beat people up.
Muhammad Ali

Debunking the "Law of Attraction"

I hope that everything you have read so far proves that the "Law of Attraction" is indeed real in the sense that anything is theoretically possible. Under these very esoteric assumptions and under extremely rarified circumstances, we can truthfully say both that:

- The "Law of Attraction" is real, and
- It is absolutely ludicrous to think that applying this "law" as its proponents would have you do can possibly yield any practical results.

A desire arises in the mind. It is satisfied immediately another comes. In the interval which separates two desires a perfect calm reigns in the mind. It is at this moment freed from all thought, love or hate. Complete peace equally reigns between two mental waves.
Sri Swami Sivananda

But what about the many testimonials of spectacular success the self-help "experts" flaunt at every opportunity? I submit that these success stories have nothing whatsoever to do with attraction and everything to do with collapse.

I fully believe that the experts and gurus who sell the "Law of Attraction" do so from a place of wanting to help others, which is the noblest possible pursuit. They are also out to make a living, and many of them do quite well financially. There is nothing wrong with that at all. It is blindingly obvious that these experts are relying on both each other and on some of the outliers of the scientific community for their information, which ultimately boils down to misunderstanding or misapplying what sciences from neurology to psychology and quantum mechanics really have to tell us. This misinformation is gathered and disseminated in good faith, of that I have no doubt; however, the fact remains that this is misinformation reveals a lack of true knowledge.

Critics and mainstream science scoff at the self-help industry, and rightly so. The sad part is that a lot of good information with the power to help many people is needlessly tainted and skewed because of its reliance on the irrelevant "Law of Attraction." The good news is that most if not all of the experts I know in the industry can gain mainstream acceptance by updating their material to reflect not attraction, but collapse (or realization, which has a more positive ring to it.)

The Law of Collapse

We know that the quantum waveform collapses whenever a measurement is taken. Physicists debate the root cause of this collapse—hence the Copenhagen, hidden variable, and many-worlds interpretations of quantum mechanics—but the bottom line is that probability does collapse into a single actuality. The Law of Collapse is thus a fundamental law of nature. We can also call it the Law of Realization because of the results of collapse. The difference depends on whether we want to focus on what is happening (collapse of the waveform) or on the results (the actuality created by the collapse). For marketing purposes, I think most people would prefer to call it the Law of Realization because "realization" sounds a lot happier than "collapse." For the sake of brevity and what I feel is slightly greater scientific fidelity, I will refer to this law simply as the Law of Collapse.

I think there is something, more important than believing: Action! The world is full of dreamers, there aren't enough who will move ahead and begin to take concrete steps to actualize their vision.
W. Clement Stone

The quantum waveform that we can calculate using the equation discovered by physicist Erwin Schrödinger is both real and central to all interpretations of quantum mechanics. It is also perfectly compatible with materialism, because it involves matter and energy whether one not one allows a role for consciousness. Dualism and idealism put consciousness into the mix, which further validates the truth of the quantum waveform. The waveform and Law of Collapse are thus compatible with all models of reality on a foundational level. One therefore need not adopt any specific belief about how the universe works in order to benefit from the Law of Collapse. Right off the bat, this is a huge advantage over the "Law of Attraction." But does it hold water?

The entrepreneur is essentially a visualizer and an actualizer. He can visualize something, and when he visualizes it he sees exactly how to make it happen.
Robert L. Schwartz

Skewing the Waveform

Self-actualized people are independent of the good opinion of others.
Wayne Dyer

Let's recap: Objects exist in a superposition of all possible states before measurement and collapse. This superposition is infinite for all practical purposes but the waveform is much higher for a very select states than virtually all other states. None of us is a "perfect" observer, and it is therefore virtually impossible to "manifest" anything. Harry Potter may be able to wave his wand and say, "Accio <object>!" to have that object immediately fly to his hand. The "Law of Attraction" would have us believe that all of us are Harry Potter incarnate, but that just is not the way of things.

In 2011, I embarked on a brutal weight-loss regimen. I might have wanted my overlarge tummy to vanish while I stuff myself with steak, lobster, and cake but, as Austin Powers said, "it just wasn't in the cards." Sure, there is a chance that I may have woken up one morning in perfect shape to fit into all of the smaller clothes I purchased as part of my visualization, but that's not the way to bet. There is also the chance that I could have kept my excellent health (my total blood cholesterol is 124) without making any changes but that too is not the way to bet. By contrast, I followed a medically supervised program of diet and weight loss supported by vitamins and FDA-approved appetite suppressing medication. The pounds literally melted away at an average rate of five pounds per week.

Before starting this program, I believed in my heart of hearts that a calorie is a calorie, and that folks like Dr. Atkins who see carbohydrates as the enemy are quacks. By everything *The Secret* says, what I eat should make no difference. Imagine my surprise when I discovered that I lost more weight during those weeks where I kept my consumption of grains, starches, and sugars to a minimum. A hundred calories of starchy baked potato is not the same thing as 100 calories of things like meat, or leafy greens.

Contentment is not the fulfillment of what you want, but the realization of how much you already have.
Unknown

This is only one example; I am sure that everyone can recount tales of discovering the falsity of some dearly held belief or truth. I am not suggesting that there is an objective reality; quantum mechanics and relativity prevent me from doing so because subjective individual realities is all we ever truly have; however, we share these individual realities with others in the framework of a lawful universe, which is why I really don't recommend defenestrating yourself anytime soon. (In other words, don't jump out a window because you think you can fly!)

My weight loss program had nothing to do with manifestation or belief. It had everything to do with taking action to quite literally skew the waveform to where the possibility of my losing weight was much higher that it could have been before. Sitting on my keister eating bonbons falls under the category of, "Yes, it's theoretically possible but don't hold your breath." where the waveform is concerned. Getting off my keister and eschewing the bonbons altered the waveform, making the possibility of my losing weight that much greater. The mere fact that I took these steps was all it takes. I didn't have to believe it would work for the waveform to skew. Yes, I believed the program could work, and I saw fabulous results. This did help tilt both the waveform and my perception of that waveform in my favor. Yes, I am happy and grateful to be feeling lighter and for the newfound energy and expanded wardrobe options I am experiencing. That also helps tilt the waveform toward maintaining my ideal weight, because it gives me incentive to stick with it despite my love of most foods sweet and starchy. This is not manifestation; it is a feedback loop where:

> *Having seen and felt the end, you have willed the means to the realization of the end..*
> Thomas Troward

1. Initial action skews the waveform toward the desired results.
2. Measurable progress (weekly weigh-ins) prove the results are on their way.
3. This progress increases my determination, happiness, gratitude, etc., which keeps me on the dietetic straight and narrow.
4. I start to believe I can do it, which gives me yet more incentive to stick with the program, and which incrementally reprograms my personal reality around food, exercise, and weight.
5. Lather, rinse, repeat.

> *There is a vast difference in some instances between what we really need and that which we think we must have, and the realization of this truth will greatly lessen the seeming discomfort in doing without.*
> William M. Peck

Over time, my personal process of building my own personal reality changed to where I now "fear" eating too much and exercising too little in the same way I "feared" eating less and exercising more. The prey instinct I describe in Chapters 2 through 4 has been reprogrammed to make me achieve my new goal of staying in shape. As an added benefit, my waveforms having to do with diabetes, cardiac problems, kidney stones, and a long litany of other ills skewed in ways that greatly reduced my odds of getting these diseases. The fact I began this program from a place of perfect health in all aspects but for the excess fat further skews my waveforms in favor of future health.

One of the best things about being an adult is the realization that you can share with your sister and still have plenty for yourself.
Betsy Cohen

In short, I have neither wished for nor manifested anything. I took action to alter my statistically probable options in favor of my desired outcome, and the waveform has done the rest, like a set of loaded dice. My daily activities alter the waveform toward my desired outcome, and my weekly weigh-ins collapse the waveform into a slimmer, sexier actuality. The Law of Collapse is affirmed.

Want a new relationship? Want money? A nice house? Vacation? Car? Pizza? Chinese food? Act! Take concrete action toward your desired outcome. Your mind and body will resist you, because that is what they are designed to do. You are a prey animal who is physically addicted to your survival instincts as directed by your core beliefs, as I discuss in depth throughout this book. Follow the four-step process of:

1. Making a goal,
2. Breaking that goal into manageable chunks,
3. Taking action on your goal one chunk at a time, and
4. Rewarding yourself for each and every chunk of progress, no matter how small...

...and you will eventually get both results and reset beliefs to boot.

You have all the reason in the world to achieve your grandest dreams. Imagination plus innovation equals realization.
Denis Waitley

You can only ever experience an outcome in accordance with the statistical odds contained in the waveform, which is why the "Law of Attraction" is worse than useless for practical purposes. By contrast, the Law of Collapse is far more powerful, because it is part of the very core of the universe's innermost workings. It is far more limited in that it promises only those odds contained in the waveform, as opposed to anything you want, any time you want it. It is far more subtle, because it achieves the real outcome by dealing with possibilities as affected by probabilities. It is also far more powerful, because it works. The universe we live could not exist without it.

You may not always get what you want. In fact, Wigner's friend guarantees that nobody ever gets anything they want. That said, you can open doors you never knew existed simply doing something, anything, to make those doors many orders of magnitude more statistically available to you than they would be otherwise.

Be happy and excited about the outcome, and you will want to persevere. Reward yourself for every little step on your journey, and you will associate your goals with pleasure instead of existential prey-based trepidation. Be grateful for every result you see and you will be happy for where you are and eager to press on, with no "promissory gratitude" for things that have not happened (and that may never happen). All of these things will tip the waveform in your favor and will help you achieve your goal. They will also make you rich beyond imagination right here, right now, no matter what your bank statement says. Loving yourself for where you are will tip the waveform in favor of finding someone to truly love you or for someone to love you more deeply.

The person with a fixed goal, a clear picture of his desire, or an ideal always before him, causes it, through repetition, to be buried deeply in his subconscious mind and is thus enabled, thanks to its generative and sustaining power, to realize his goal in a minimum of time and with a minimum of physical effort. Just pursue the thought unceasingly. Step by step you will achieve realization, for all your faculties and powers become directed to that end.
Claude M. Bristol

The "Law of Attraction" is all about the destination. In this lifetime, the only real destination is death, which may be the end of the line (although I don't think so, hence *The Divine Savage*). Life—and the Law of Collapse/Law of Realization—are all about the journey. Make your journey a good one!

Beliefs Don't Matter (Much)

In this chapter, I assert that beliefs don't matter, that action is all that matters. I have also asserted throughout this book that only reason I am sitting at my computer typing is because I believe I am. I further assert that the brain cannot tell dream from reality and that there may be no difference. It may seem like I am contradicting myself, but that is not the case. Beliefs do matter, because they shape our personal realities. But think about it: Have you ever done something that you truly believed to be absolutely impossible? I submit that the only way that you will jump out a window is if you are either convinced to the core of your being that you can fly (again, not recommended), or if you want to end this lifetime and believe that a long fall and resulting sudden stop is the way to do it. I would never have begun my weight loss program if some tiny part of me did not think it possible—however unlikely—for it to work. Belief on a conscious or (usually) unconscious level is always at work. Practically, however, it is the action that matters. The Catch-22 of being unable to change beliefs without results and unable to get results without belief is thus avoided... for the simple reason that we are not perfect.

Accept no Limitations

More powerful than all the success slogans ever penned by human hand is the realization for every man that he has but one boss. That boss is the man—he—himself.
Gabriel Heatter

Many self-help experts know that sitting around wishing it so does not do much to actually make it so, which is why they shield themselves from responsibility (and potential liability) by couching the "Law of Attraction" in terms of taking action. For example, author Alexander Kjerulf says that, "I believe that the 'Law of Attraction' is very real. I've used it on any number of occasions. It works." Kjerulf then goes on to say that, "Changing your thinking changes nothing out there, in the vast universe surrounding you. It changes something inside of you. Changing your perception, your focus, your emotions and your thinking from negative to positive (from what you lack to what you want) has an enormous effect on your motivation, energy, and creativity and that's why you will then be more efficient working towards your goals." In other words, if you want something, do something! This sounds a lot less like the "Law of Attraction" found in *The Secret* and its many spin-offs and copycats and a lot more like—you guessed it—the Law of Collapse.

Living the Real Secret

Success is the progressive realization of a worthy goal or ideal.
Earl Nightingale

If you want to change something in your life, take action. Use the processes contained in this book to find any creative way you can to skew the waveform for whatever it is you want in your favor. For example:

- I love sailing, and am actively on the market for a sailboat. My dream boat is a 50-foot Hunter sloop, but I don't have a spare half-million dollars lying around. I can buy a much cheaper used sailboat, and then invest significant time, effort, and money into fixing it up and maintaining it... or I can pay a very modest fee for a stake in a Hunter that is very close to my ultimate goal and have all the maintenance, insurance, cleaning, dock fees, registration, etc. taken care of for me at a cost that is very competitive compared to keeping a "cheaper" used boat shipshape. It is also a great way for me to refine my sailing skills and prepare for the day I can afford my own dream boat on which to retire to the South Seas with Jennifer.

- I love flying. Even the worst day goes down as a good one in my memory when I get a chance to fly. I can't afford the cost of buying, maintaining, inspecting, and insuring a plane.... but my local flight clubs have whole fleets of airplanes at my disposal whenever I want, and I get a much greater variety of airplanes to choose from instead of being locked into only one bird.

The object of living is work, experience, happiness. There is joy in work. All that money can do is buy us someone else's work in exchange for our own. There is no happiness except in the realization that we have accomplished something.
Henry Ford

The list goes on and on. The bottom line is that if you want to do something, then there is probably a creative way to do it. It may not meet your exact desires, but a stepping stone is both progress and a great way to make sure that you really do what you think you want. It also a great way to protect your future investment when and if you make it. When I finally buy my own Hunter, I will know all there is to know about Hunter boats, their strengths, weaknesses, how to maintain and operate them, etc. than I ever could by simply buying a new one. That insight will save me time and money down the road and could even save my life.

Be happy about the many blessings you do have in life, no matter what changes you want to make. Be grateful for all you do have no matter how little of it you have. Love yourself for all your flaws and foibles. Nothing in life is ever inherently good or bad, unless and until you make it so by choice. Learn to consciously decide what is good and what is not good, instead of relying on your subconscious processing to do it for you. Choose to see the good.

Do this and you will be healthier, happier, and richer than you are now, no matter what happens or doesn't happen. Do this, and your waveform will stretch as needed to help you along the way. The Law of Collapse guarantees that something will happen. How you use it has a profound effect on what happens.

When one is rising, standing, walking, doing something, stopping, one should constantly concentrate one's mind on the act and the doing of it, not one ones' relation to the act or its character or value.
Ashvaghosha

Don't "attract" your dream reality. Collapse it. Realize it! This is the only way to make what you want come to pass. There is no other way.

Chapter 24

The Sky is Not the Limit

Congratulations! You are now free to achieve and surpass your life's goals, and to have all of the abundant financial, emotional, and spiritual wealth that you deserve for being alive and true to your dharma. Remember how your brain threw up obstacles whenever you tried to deviate from your beliefs? Well, it's still going to do that; however, this time it's a good thing... if you have taken all of the lessons in this book to heart, that is.

Success Begets Success

Accept the challenges, so you may feel the exhilaration of victory.
George S. Patton

You've been 100% successful at following your old programming, which should give you every reason to believe that you can be 100% successful at following your new programming. Welcome to your new life as an enlightened savage! How does it feel? I listen to Internet radio. As I was editing this paragraph, a song came on about rebirth. Funny, because that is exactly what's happened to your programming! Your old life, with its limitations and negativity, is over.

Had anyone told you at the beginning of *The Enlightened Savage* program that you could erase a lifetime of self-destructive beliefs, purge yourself of negative emotions, free yourself from nagging negativity, and cast aside non-productive expectations in hours, what would you have said? Having been through this process and become an enlightened savage, what do you think now?

Reflections on Change

The processes you experienced in chapters 11 through 14 were developed and copyrighted by Dr. Morty Lefkoe's Decision Maker Institute. Dr. Lefkoe has helped thousands of people overcome negative programming. He can help you too. If you need personalized assistance with any of the processes, or are experiencing any difficulty with them, I urge you to contact the Decision Maker Institute. Morty and his staff are always ready to help. If you need more help with letting go of attachments and clearing negative emotions and beliefs, then I can't recommend Jim Britt's *Power of Letting Go* and *Rings of Truth* enough. Jim is the real deal.

The way of the creative works through change and transformation, so that each thing receives its true nature and destiny and comes into permanent accord with the great harmony: this is what furthers and what perseveres.
I Ching

Did the processes described in *The Enlightened Savage* work for you? By now, you probably know the answer: They worked as well as you made them work, because you alone build your own personal reality. They worked for me; I therefore know that they can completely alter your life forever. The moment my programming changed, I felt a sense of euphoria that's still with me years later, and that's getting stronger all the time. Oh sure, life remains challenging. But I now know I can meet those challenges head on. How? Here's my story; judge for yourself.

In late 2008, I found myself divorced and sans income with a pile of debt and no prospects in sight. Going to my family and friends for help was one of the most humbling, humiliating things I've ever done. Every dollar I spent, every bite I took, reminded me that I was depending on others. That's when I finally grasped the extent to which my beliefs were driving me into the ground.

Remember my microscope story from Chapter 4? That's how strong core beliefs are. I spent money like there was no tomorrow because I couldn't stand the idea of being deprived. This habit resulted in deprivation, because I didn't know how to manage what I had. I now keep a microscope on my desk as a constant reminder.

Despite this, I remained convinced that my situation was somehow a blessing. You see, my professional life had been a series of accidents, and I was convinced that the Universe was trying to point me in a new direction. Writing is my passion. Some of my work allowed me to meet many new authors. I quickly learned that sales and marketing are their biggest challenges. I founded Dawnstar Books and created videos to teach authors critical business skills in artist-friendly terms. My ideas were sound, but I knew that I needed a big endorsement from a big name. So what did I do?

Some years before, I had decided I'd ask the world's undisputed marketing champion, Mr. Jay Conrad Levinson, to endorse me. Guess what: Not only did he endorse my products, he opened doors I never knew existed—and *The Enlightened Savage* is only one of them. I met Jim Britt thanks to connections made through the Guerrilla Marketing Association and have other business opportunities as well. You see, my past jobs and businesses all revolved around education and helping others. I was on the right path, but not moving with a clear purpose.

What a man can be, he must be. This need we call self-actualization.
Abraham H. Maslow

I am fixing that problem, and stand before you today on the path to more personal and financial success than I ever dreamed possible. Why? Because I changed my programming! The Universe would not have blessed me with Jay's support if I was not able and ready to receive it. It took me being knocked around a little to get that through my head. It is my sincerest hope that you, too, will make and live whatever changes you need. I know you can. Why? Because if I can do it, then so can you. So can anyone.

Yes, it really is that easy—so easy in fact, that my single largest difficulty was getting past the sneaking suspicion that I was doing something wrong, or that the other shoe would drop. Well something did go wrong. Plenty of things went wrong, in fact. Why? Because I was living in a state of fear and unconsciously choosing to see challenges as negative. I marvel today at how I calmly face challenges far greater than past challenges that used to reduce me to tears. Do Dr. Morty Lefkoe's processes work? How about Caterina Rando's *Success With Ease* program or T. Harv Eker's *Millionaire Mind Intensive* or Jim Britt's *Power of Letting Go*? They are working for me, and they can and will work for you—if you want them to.

We've talked a lot about cause and effect. We've also discussed the ETEAR process of realization, where emotions lead to core beliefs, which drive actions, which generate our realities. We know that our physical world only represents the effects. We've just spent a fair amount of time changing your programming to build a much better personal reality around a framework of non-attachment and conscious choice, to ensure that you actively build the personal reality you want.

I have already promised you that the changes you are about to experience will be miraculous, if you say, "Thanks for the step up!" whenever obstacles confront you. The prison walls our brains erect around us are only illusions. There are no guards, no fences, no bars, no nothing.

I've told you my story. What about your story?. Let's start with your parents. How did their beliefs reveal themselves? Did your old beliefs and actions mirror your parents' or did you rebel? I have a friend whose parents are very well off. Meanwhile, he's perennially unemployed and ticked off with the world. Something in his childhood affected him so seriously that he vowed never to follow his parent's footsteps.

You have all the reason in the world to achieve your grandest dreams. Imagination plus innovation equals realization.
Denis Waitley

Another friend of mine was struggling to rebuild her music career. Her parents spoiled her rotten until she hit her teens. Then they used her passion for music against her by threatening to give away her instruments, saying that she was a useless, worthless piece of trash. She confided that she felt the urge to call her father for approval every time she accomplished something. Why? She discovered that her programming made her happiness dependent on other people, because she wasn't good enough to seek it on her own. Her desire to call her father stemmed not from seeking approval but disapproval!

Acting on her own contradicted her programming; her brain fixed that by driving her back to the source of her programming while tricking her into thinking that she was seeking approval, because no one consciously seeks disdain. How's that for a computer virus? Her life changed the moment we exposed her beliefs. In just a few weeks, she did more for her career than in the previous ten years combined.

A third friend wanted to found an intentional community of people living and working for common goals. His core beliefs had him doing clerk and temp work that's about as far from community leadership as one can get. A few weeks of coaching, and he started getting his core story together. He moved into an existing community and began actively moving towards his dharma. That's how powerful core beliefs are, and that's how easy it is to change them and make real progress towards the things that matter.

He who rejects change is the architect of decay. The only human institution which rejects progress is the cemetery.
Harold Wilson

You know that your physical world is nothing more than the sum of your three invisible worlds: Mental, emotional, and spiritual. You must balance these three elements for maximum success. Do this by always placing yourself and your needs ahead of anyone and anything else. What are your needs? Anything and everything that will help you achieve your life's goals.

Whatever you put energy into grows. Put two identical plants side by side. Water and feed one and abandon the other. What happens? Case closed.

You Come First

It is better to conquer yourself than to win a thousand battles. Then the victory is yours. It cannot be taken from you, not by angels or by demons, heaven or hell.
Buddha

Pay yourself first, and pay yourself right off the top before paying for anything else. If you're worried about not having enough, you won't, because that's the personal reality you're creating for yourself. Pay yourself first, and there will always be enough left over for the other bills.

Pay your bills first and there will never be enough left for you. Don't get me wrong: I not telling you to spend all your money on wine, women, and song. What I am saying is that you are the most important person in your life. Treat yourself that way, because your personal reality cannot exist without you—not the other way around!

Physical Health

Disease is an experience of a so-called mortal mind. It is fear made manifest on the body.
Mary Baker Eddy

Take care of your physical quadrant. I've started exercising a few times per week, and my energy levels and sense of well-being and happiness have skyrocketed. Get your blood flowing! Aerobics, weights, swimming, hiking, rowing, dancing, calisthenics, tai chi, yoga—it doesn't matter.

While we're at it, tidy up your home and office. Get rid of the dishes and dust bunnies. Make your environment as pleasant as possible to be in. You may have a very modest home, or may even be sleeping in your car. Hey, that's OK. Make your environment as clean and inviting as possible. Make it reflect the fact that you deserve the best! Whatever you put energy into grows.

Emotional Health

Take care of your emotional quadrant. No negative self-talk! The moment you hear yourself saying, or even thinking, something negative about yourself, stop! Replace it with something positive.

The key to success is to keep growing in all areas of life—mental, emotional, spiritual, as well as physical.
Julius Erving

Spiritual and Mental Health

Take care of your spiritual quadrant. However you choose to express your spirituality and give thanks to this loving Universe we live in is perfectly OK. Last, but certainly not least, look to your mental quadrant. Remember, if you're not growing, you're dying!

The great awareness comes slowly, piece by piece. The path of spiritual growth is a path of lifelong learning. The experience of spiritual power is basically a joyful one.
M. Scott Peck

This is Just the Beginning

Becoming an enlightened savage is giving you one heck of a great start. But don't stop here! *The Enlightened Savage* is the first of three distinct "layers" in the *Savage* series. I encourage you to read *The Natural Savage* to learn the science behind the processes in this book. *The Natural Savage* explains how all we ever do in this life boils down to the following six core functions: predator avoidance, group status, food, shelter, reproduction, and death. From there, I am writing six more Savage books that explain each core function in detail... and how to get the absolute most from your prey brain. Of these, *The Divine Savage* peels away the onion-like layers of shared and personal reality to reveal the truth of what happens when we die—a truth that is beyond anything most people can imagine.

We continue to shape our personality all our life. If we knew ourselves perfectly, we should die.
Albert Camus

You've seen how all of the energy you put into your old beliefs grew and expanded until it ruled your entire life. Instead of focusing on these old negative beliefs, focus on your life's goals and what it will take to get there. Just doing that will make the kind of opportunities and results that will propel you forward possible.

Country music star Deana Carter summarizes my philosophy beautifully:

> *Every single spirit deserves to be happy, and I'm gonna remind them. It's what keeps me in this and it's why I want to be able to do the splits when I'm 40. Because to not pursue happiness would be to compromise your ethics, your vigor for life, and life is what you deserve.*

Beautifully said, Deana!

Commit to Yourself

My satisfaction comes from my commitment to advancing a better world.
Faye Wattleton

The time is now. Decide how you will place yourself at the top of your priority list, and how you will tend and nurture each of your four quadrants. You are making truly amazing progress. The best is by far yet to come. Does that feel fantastic or amazing? Expect success. Plan for it. Then go make it happen. You deserve it. You deserve it! YOU DESERVE IT!!!

Commit to Success

You get whatever accomplishment you are willing to declare.
Mal Pancoast

If you read this book with a truly open mind, and allowed me to show you a new way of living your life, then you have the raw ingredients you need to begin achieving great success. You know how to do it. But will you do it?

Commitment is everything. Nothing happens until you make it happen; everything that happens occurs because you realize it. I have given you the tools and the know-how, but I can't force you to take action. Well, maybe I could, but you wouldn't be acting of your own free will.

Magical You

Magic is believing in yourself, if you can do that, you can make anything happen.
Foka Gomez

The dictionary defines *magic* as, "the art that purports to control or forecast natural events, effects, or forces by invoking the supernatural." Most people think that everything they see, hear, touch, taste, and smell causes their personal reality. You now know that the opposite is, in fact, true. Creating your own personal reality is therefore pure magic, making you a real-life wizard. That's right: You are the most powerful wizard imaginable, for you can conjure up whatever destiny you can imagine, within the limits of physical laws that we explored in Chapter 23. What a waste to not use your almost unimaginable power!

You may get stuck at times, or not know where you're going. That's OK. Just remember to surrender your need for control and to exercise your ability to consciously choose how to interpret everything in your life. You'll soon find yourself back on track.

A New Enlightened Savage

The moment of enlightenment is when a person's dreams of possibilities become images of probabilities.
Vic Braden

Let's take one final look at the amazing things you've just accomplished: We started out by setting some goals. Next, we eliminated your negative programming, using both on- the-spot processes as a sort of virtual surgery and ongoing cognitive processes to speed your recovery and adaptation to your new beliefs and expectations. I showed you some methods to help find your mission in life, and guided you through mapping out your goals and how and when you will achieve them. Now, it's up to you to either build a new personal reality filled with boundless possibilities or plop back on the sofa with a bag of chips and the remote. It's up to you. It has always been up to you.

This concludes *The Enlightened Savage.* I hope this book will be your gateway to becoming the person you are meant to be. My dharma is to help you become the person you were born to be in both your business and personal lives.

I look forward to meeting you again in future *Savage* books and to sharing the joy that comes from knowing that the sky is not the limit. Thank you for reading. Good bye for now.

So many of our dreams at first seem impossible, then they seem improbable, and then when we summon the will, they soon become inevitable.
Christopher Reeve

Resources

I strongly recommend the following additional books and other resources. Each of these has provided tremendous value in my life, and I hope they can do the same for you. These are just a small sampling of the many fantastic resources that discuss the concepts presented in this book, and are not intended as a complete bibliography. I present them in the hopes that you will be inspired to keep on learning and growing.

Books

- *Callings* - Gregg Levoy
- *Depression: Types, Concepts, & Theories* - P. Gilbert
- *Do This. Get Rich! for Entrepreneurs* - Jim Britt
- *Do This. Get Rich! for Network Marketers* - Jim Britt
- *Emyth Mastery* - Michael Gerber
- *Guerrilla Marketing* - Jay Conrad Levinson
- *Instinct for Dragons* - David E. Jones
- *Molecules of Emotion* - Dr. Candace Pert
- *Nickeled & Dimed* - Barbara Ehrenreich
- *Paths From Science Towards God* - Arthur Peacocke
- *Rings of Truth* - Jim Britt
- *Power of Letting Go* - Jim Britt
- *Science vs. Religion* - David S. Turell
- *The Emyth Revisited* - Michael Gerber
- *The Field* - Lynne McTaggart
- *The Nature of Consciousness* - Jerry Wheatley
- *The Natural Savage* - Anthony Hernandez
- *The Self-Aware Universe* - Amit Goswami
- *Subordination and Defeat* - Leon Sloman & Paul Gilbert (ed.)

- *Tree of Origin: What Primate Behavior Can Tell Us About Human Social Evolution* - Frans B. M. De Waal (ed.)
- *Way of the Guerrilla* - Jay Conrad Levinson

Articles

- *Gene Mutation of Jaw Linked to Brain Evolution* - Washington Post, Nature, et. al

Web Sites

- Decision Maker Institute - www.decisionmaker.com
- Guerrilla Marketing - www.guerrillamarketing.com
- Power of Letting Go (Jim Britt) - www.jimbritt.com
- Peak Potentials (T. Harv Eker) - www.peakpotentials.com
- Success with Ease (Caterina Rando) - www.caterinar.com

Movies

- *The Elegant Universe*
- *What the Bleep Do We Know?*

Appendix

14-Week Follow Up Plan

Go online to www.dawnstarbooks.com/tes/materials to obtain The Enlightened Savage Companion, which contains a more detailed follow-up plan and a complete Goals Matrix.

Chapter 19 stressed the importance of purchasing and using a day planner to help you track your appointments, errands, and expenses, and (more importantly) keep you on track toward accomplishing your goals. This Appendix consists of one mini-planner for each of the 14 weeks after you complete *The Enlightened Savage.* For a more detailed plan, visit www.dawnstarbooks.com/tes/materials to download a copy of the Enlightened Savage Companion.

The Importance of Following Up

It is very important that you set goals and move towards them using the success strategies from Chapter 19. Doing this will allow your new programming to take root and become part of your brain's internal pathways, in the same way that your old negative programming was part of those same pathways. Stop now and you risk reverting to your old beliefs. Press forward, and you'll leave your old core beliefs behind forever.

How it Works

I use a very simple four-step process with my coaching clients to help them achieve and surpass their goals. This process is designed to gradually change and upgrade mental programming in much the same way as the exercises you experienced in this program. I have seen people experience amazing transformations using the approach I am about to present to you.

Define Your Goal

What would you like to accomplish during the next fourteen weeks? Imagine yourself in San Francisco about to walk to New York, about 3,000 miles.

Pick Easy Milestones

Set off from San Francisco with New York in mind, and you'll arrive in Oakland, California (about 10 miles from San Francisco) hot, tired, sweaty, and ready to call the whole thing off.

I advise all of my clients to forget about the destination and to focus only on the next step. Worrying about where you've been is pointless, because you can't change that. Worrying about what's coming two steps from now risks having you trip yourself up, because your eyes aren't looking where you're going. The idea is to take small steps, and to celebrate each step for the real and measurable progress that it represents.

Taking easy steps accomplishes many things, including:

- Monumental tasks become easy and fun.
- You can celebrate lots of little accomplishments instead of bemoaning the big one that got away.
- If you get uncomfortable (what I call "hitting the Rocky Mountains"), the point where the gopher realizes it's eased halfway out of its hole and wants to jump back in), take smaller steps.
- Like a loaded freight train, you will need some time to get up to speed. Slow gentle acceleration is the best way to ramp up.
- Like a loaded freight train, once you're up to speed, you'll have lots of momentum and won't want to stop-which is, of course, the entire point of this exercise.

Perform the Task

At some point, you must actually take the step, if you are to move forward towards your goal. Breaking your journey down into many small easy steps, and focusing only on your next step, makes the journey as easy as possible. All that's left is for you to move forward.

Set a clear destination for the step that you can see, and then accomplish that step with as much ease as possible. Define criteria for successful completion, and give yourself a deadline. That done... go for it.

Reward Yourself

Here's the good part: The whole purpose behind *The Enlightened Savage* is to bring you to a place where you are associating joy with living your dharma and accomplishing your goals. The best way to do that is to define a reward for each step of each journey you take. If you accomplish your step on time, and to the success criteria you defined, you get the reward. It's that simple. If you don't accomplish the step on time or to spec, no problem; you just don't get the reward. Punishment is never an option, because you've had enough pain in your life. In fact, pain is what drove you to *The Enlightened Savage* in the first place.

Every reward you give yourself must meet the following three criteria in order to have the maximum effect:

- **Originality:** Your rewards must be creative. If you go out for dinner all the time, don't reward yourself by going out to dinner yet again. Be creative! The reward need not cost much or require much time or effort. It just needs to be creative.
- **Meaning:** The reward must have some deeper meaning for you. For example, you might decide to take time off and go see the sunset to reconnect with nature and the beauty of this world. You might decide to get a spa treatment to experience what it's like to receive, because your life involves so much giving. The possibilities are endless.
- **Timeliness:** Always reward yourself as soon as possible after finishing each step. You've experienced enough

delayed gratification in life. It's high time your mind saw a clear and unmistakable correlation between fulfilling its dharma and getting the sweet treats.

Easy steps with a reward at the end of each one. Does this sound great or great? The best part is that it works.

Get Support

It's one thing to read this process and have the tools in front of you. Having someone to support and encourage you at every step is a whole new experience.

Week One

- **Success Strategy:** Choose joy

Find ways to experience joy in everything you do. Your new programming is very young and will face many challenges as your brain tries to revert to its old ways. Make sure to reward yourself every time you successfully complete a task on time.

Week Two

- **Success Strategy:** Live on purpose

On a practical level, there are no such thing as accidents, coincidence, or randomness. You were born and are on this planet for a reason and have a very specific mission to accomplish. How can you tailor your life and the things you do every day to fulfill that mission or dharma?

Week Three

- **Success Strategy:** Acknowledge others often

No person is an island. Each of us depends on many other people for the many essentials of daily living. Take a few moments every day to acknowledge the efforts of the people who help you in so many ways. Don't overdo it; simple sincerity is always best.

Week Four

- **Success Strategy:** Ask for what you want

It is impossible for you to get everything you want by yourself. Unfortunately, society conditions us to believe that asking for what we want is selfish. Be true to yourself. If you want something, never be afraid to ask for it. You'll be amazed at how often you get what you want.

Week Five

- **Success Strategy:** Be willing to be uncomfortable

Are you feeling more than a little uncomfortable with your new programming? By now, your brain is probably screaming at you to get back in your burrow. That's OK. You are making great progress. Allow the discomfort to exist and know that it means you're on the correct path.

Week Six

- **Success Strategy:** Explore new possibilities

By now, you're starting to make some real progress towards your goals and are hopefully starting to experience the thrill of moving forward faster and easier than ever before. As you become more comfortable with achieving your goals, what new horizons can you see?

Week Seven

- **Success Strategy:** Maintain a positive disposition

A glass filled to 50% capacity is either half full or half empty. Which is it? Whichever you want it to be. You're probably used to seeing this proverbial glass as being half empty. What if you could choose to view the same glass as being half full? Guess what: You can.

Week Eight

- **Success Strategy:** Take small actions

Grabbing a gopher and yanking it out of its burrow is the best way to make sure it never wants to emerge again. Your brain works the same way. Break your large goals into small easy steps. Little by little, you'll go far. Remember to reward yourself for each completed step.

Week Nine

- **Success Strategy:** Openly express your gratitude

As your new programming takes hold and you find yourself moving closer to living your dharma, pause to think about the many people helping you on your way. Make sure to thank each of these people for their contribution, no matter how small it might seem now.

Week Ten

- **Success Strategy:** Have some fun

All effort and no play is the best way to revert to your old negative programming. Your life and your goals exist to serve you, not the other way around. Make very sure to set aside plenty of time for just having fun. If this is part of your goals, fine. If not, that's just as OK.

Week Eleven

- **Success Strategy:** Smile, laugh, and love more

I hope that your life is filled with more joy and abundance than ever before by this point. How does it feel? Joy is contagious. If you're feeling more joyful, express it! Smile and laugh as often as possible. Make sure you let those closest to you know how much you love them.

Week Twelve

- **Success Strategy:** Look for the ease

Remember that "work" is defined as results, not as effort. There is no such thing as working hard, just hard struggle. Focus on struggle and you'll be sure to experience lots of struggle. Focus on results and your life will get much easier.

Week Thirteen

- **Success Strategy:** Always expect success

Never forget that you and you alone create your personal reality. This means that you can choose to realize the success you seek and deserve in your business and personal lives. Go into every situation and task expecting success and success will find you far more often than not.

Week Fourteen

- **Success Strategy:** You have the power

You cannot read this often enough: You create your own personal reality. This means that you have the ultimate power over and responsibility for your life. Accept and own this power and use it to create the personal reality you seek, the reality you were born to have.

CPSIA information can be obtained
at www.ICGtesting.com
Printed in the USA
LVOW05s0225200617
538609LV00028B/858/P

9 780985 579333